D1112318

From EDI to Electronic Commerce

Titles of Related Interest

Fatah
ELECTRONIC MAIL SYSTEMS: A NETWORK MANAGER'S GUIDE, 0-07-020056-4

Hardman
NETWORK FAXING: CHOOSING AND USING YOUR COMPUTER-BASED FAX, 0-07-026267-5

Hertzoff
AT&T GLOBAL MESSAGING: THE EASYLINK COMPLETE REFERENCE, 0-07-028459-8

Hogbin
INVESTING IN INFORMATION TECHNOLOGY: MANAGING THE DECISION-MAKING PROCESS, 0-07-707757-1

Jayachandra
REENGINEERING THE NETWORKED ENTERPRISE, 0-07-032017-9

McClain
OLTP HANDBOOK, 0-07-044985-6

Minoli
IMAGING IN CORPORATE ENVIRONMENTS, 0-07-042588-4

Minoli
ANALYZING OUTSOURCING: REENGINEERING INFORMATION AND COMMUNICATION SYSTEMS, 0-07-042593-0

Radicati
ELECTRONIC MAIL: AN INTRODUCTION TO X.400 MESSAGE HANDLING STANDARDS, 0-07-051104-7

Spohn
DATA NETWORK DESIGN, 0-07-060360-X

Wang
TECHNO VISION: THE EXECUTIVE'S GUIDE TO UNDERSTANDING AND MANAGING INFORMATION TECHNOLOGY, 0-07-068155-4

From EDI to Electronic Commerce

A Business Initiative

Phyllis K. Sokol

McGraw-Hill, Inc.

New York San Francisco Washington, D.C. Auckland Bogotá
Caracas Lisbon London Madrid Mexico City Milan
Montreal New Delhi San Juan Singapore
Sydney Tokyo Toronto

Library of Congress Cataloging-in-Publication Data

Sokol, Phyllis K.
From EDI to electronic commerce : a business initiative / Phyllis K. Sokol.
p. cm.
Includes index.
ISBN 0-07-059512-7
1. Electronic data interchange. 2. Business—Data processing. 3. United States—Commerce—Data processing. I. Title.
HF5548.33.S653 1994
005.74—dc20 94-41344
CIP

This book is a revision of *EDI: The Competitive Edge* published by McGraw-Hill in 1989.

6 7 8 9 0 DOC/DOC 9 0 9 8

ISBN 0-07-059512-7

The sponsoring editor for this book was Marjorie Spencer, the editing supervisor was Mitsy Kovacs, and the production supervisor was Suzanne W. Rapcavage. It was set in Palatino by McGraw-Hill's Professional Book Group composition unit.

Printed and bound by R. R. Donnelley & Sons Company.

This book is printed on recycled, acid-free paper containing a minimum of 50% recycled de-inked fiber.

To Ray Greenspan (my mom),
Scott, Bruce, and Eric Sokol (my sons),
and Larry Shelley (my friend)

****

It's never too late to change ... and it's never too early to start thinking about it.

Contents

Preface

This book is a major revision to my previous book called *EDI: The Competitive Edge*. I started that preface with the following sentence: "By every indicator, EDI, electronic data interchange, is taking the business community by storm." In the next paragraph I predicted: "By 1993, 70% of all U.S. companies will be making significant use of EDI." Neither statement proved to be accurate. While many companies had begun to use EDI and many more have since joined the ranks, EDI is not taking the world by storm. Even in those companies that are doing EDI, transmitting and receiving data electronically, it is rarely the primary mode of doing business. Aside from just a handful of noteworthy cases of companies that have used their vast influence and clout with their trading partner base to convince them all to implement EDI for at least one application, other applications have usually not followed suit and are still being handled through a combination of system and manual means.

To further reduce my credibility, I will tell you that I blithely continued in my previous preface to remark that "Those that implement early will realize a competitive edge over their more traditional competitors." If there's any one thing that I have learned since that preface in 1990, it is that EDI does not *give* you a competitive edge! What it does do is *position* you to gain competitive advantage. Electronic data interchange is certainly a first step to allow you to generate and transmit machine-readable files that are readily acceptable to your trading partner base. Likewise it would be almost impossible for you to be an effective receiver and handler of machine-readable data without the standardized formats that have become the trademark of EDI. In short, without EDI it would be difficult, if not impossible, to provide the high service levels and additional information that are fast becoming the norm. The problem is that you can easily do EDI and not get any of these advantages. In fact, unless implemented properly and integrated into your day-to-day business procedures, EDI can add overhead to your already straining work force and routines instead of saving time and money.

This brings us to the follow-on comments in the previous preface. There I said,

> Because of its inter-company aspects, electronic data interchange has been assumed to be a data processing and communications project and has been assigned to programmers, analysts, and communication specialists. This is not the case. While computers and software programs are needed to support EDI and communication of business transactions, EDI is really a new way of conducting business. As such, it should be the province of functional business managers with strong support from their top level managers to evaluate potential benefits of EDI, develop new business procedures for the electronic environment and develop business specifications for system enhancements and development.

My second biggest lesson since 1990 has been that this statement was not only true then but should still be shouted loud and strong from the rooftops today. I will venture to say I am not aware of a single company that is successful in EDI today who does not openly credit business involvement and top management support as the two main reasons for their success in EDI. Even those that have used clout to influence their trading partner base have been able to do so only because their chief executive officers have stood strongly behind the initiative and have made that clear by signing EDI-related correspondence to those business partners.

This book, like the previous one, is intended for functional business people. Its focus is on business issues, integration of EDI data into the system environment, cost-benefit analysis, and planning and implementation. However, in addition, I stress what I call "rethinking" the way business is accomplished today, and I offer a detailed, step-by-step approach to discover how business is being accomplished today, to evaluate the effectiveness of that approach, to revise and streamline the flow of business information, and to automate manual procedures. In today's jargon, this is what is popularly referred to as "reengineering." I strongly believe that this is the missing link between mere implementation of EDI and realizing savings from implementation of EDI. In addition, while the primary focus of this book remains on EDI with its machine-readable business transactions, I also recognize that EDI does not satisfy the full intercompany messaging requirements of business entities today. So, the secondary focus is on electronic commerce and the sharing of information using many different electronic means such as E mail, fax, down-loads of files, CAD/CAM drawings, bar codes, electronic catalogs, and others.

In this book, I have added a chapter on who should get involved during the planning, implementation, and ongoing support phases of your EDI program. I believe that this will be very helpful to those of you just entering the EDI arena. You'll notice that during planning, I'm recommending fairly high-level, strategic planners. When these people buy into the program, two very important things happen. One, the program is designed from the outset to support the strategic objectives of your organization, the mark of a sure winner. Two, it actually happens! Rarely does the program flounder that is supported by high-level executives. You'll also notice that as you move on to the implementation stage, the cast of characters changes. The people we want to implement the program are those that have the most to gain when the program is successful. These are typically the mid-level managers, with their fixed budgets, quotas, and head count, as well as outside pressure from trading partners, competitors, and the domestic and global marketplaces. Finally, for ongoing support we need the continued time and effort by the workers, both functional business people and IS staff. Electronic data interchange is not a perpetual motion machine, moving along forever with one small push to start it off. While most of the support activity is shouldered by IS people, don't think that the business people are completely out of the loop. Much of the success of EDI is in knowing what EDI is and what it isn't. What *EDI isn't* is a replacement of the business information and ongoing business relationship between trading partners. Be sure your business people remain the primary liaison between your organization and your business partners and continue to be the ones with whom business information is discussed.

While many of the chapters in this book cover the same topics as in my previous book, they are all substantially, if not totally, rewritten to reflect what I have learned since I wrote the previous book and to present the material in a more practical and usable way.

In addition, I have substantially revised the industry chapters. For the healthcare and retail/grocery chapters, the material reflects a more abbreviated version of the beginnings of EDI and deals in greater detail with what is happening today and where we anticipate it is going. For example, in the retail industry chapter there are in-depth discussions of both efficient consumer response for the grocery segment and quick response for the mass merchandiser segment. This not only broadens the perspective of this book from EDI to electronic commerce but addresses the business issues related to both initiatives. In the health care industry, the focus has also changed dramatically. In my previous book, the big news was EDI between pharmaceutical manaufacturers and wholesalers. In this book I have presented a broad view of the health care industry as a whole and have discussed what is happening with insurance companies, health care providers, hospitals, drug chains, and pharmacies, as well as the manufacturers and wholesalers. These industry chapters are intended to provide for you a picture of what is possible in EDI and electronic commerce. You will read about many different applications that were selected based on the business needs of the industry. It is hoped that they will spark your imagination of what is possible in your industry.

Finally, the last chapter reflects futures in electronic commerce. Here I describe various technologies that are just beginning to be thought of as part of electronic commerce and others that will potentially be used in the future.

To write this book, I continued to draw from my educational and work experiences. As data-processing manager of a corporate EDI project, I designed and developed the software needed to process, translate, and control EDI transactions for a trading community of almost 200 trading partners. As an active member of the ANSIX12 committee since 1983, I participated in the maintenance and public relations subcommittees and in several task groups. As design consultant and sales support with two EDI VANS, I designed the EDI management software necessary to translate and track EDI data going through the network. As an independent consultant I made contributions to the planning and implementation of EDI in several major corporations. Today, as Director of Educational Services for Sterling Software's Network Services Division, I remain in the mainstream of EDI growth. Our educational group both develops and teaches EDI courses for novices as well as for experienced EDI people, for technical as well as for business interests. In addition, in December of 1992 we introduced what has become an important annual event in EDI, the EDI/EC Coordinators' Conference. My interaction with the more than 300 attendees each year has given me a clear picture of where companies are today and where they are headed in the future. It was as a result of this group that I realized that companies are beginning to broaden their perspective from EDI to electronic commerce and that a new book was necessary.

I know that my varied experiences have been invaluable to me in understanding what is needed to implement EDI and electronic commerce and to gain real savings from them. I hope that the presentation of that perspective in a simple, down-to-earth manner will be valuable to you as well.

Acknowledgments

I would like to give special thanks to Greg Swanziger, Jim Sykes, Robert Moore, Tom Sample, Jeannine Hunter, Barb Chapin, Bob O'Malley, Peter Meuller, and Lee Bonneau for sharing with me their in-depth knowledge of their respective companies, industries, and fields of expertise; to John Stelzer, Steve Love, Dan Davis, Hazel Boss, Tim Rose, Jack Hyer, Dave Dodge, Jim Crossley, Rob Humphrey, and Sheila Goad, work colleagues with whom I've discussed many of the concepts of this book; and to Larry Shelley for offering support and encouragement for this and other ventures over the past 5 years.

Phyllis K. Sokol

1

Setting the Stage for EDI

Objectives

- Demonstrate how the current competitive worldwide business environment has increased the need to cut costs throughout the organization.
- Show how just-in-time (JIT) manufacturing and ordering schedules can cut internal costs but can be maintained only with JIT information throughout the organization.
- Show how the move toward quality and higher customer service levels has impacted traditional business practices and procedures.
- Describe how the trend from a centralized environment to a distributed one has required the need for more timely and localized information.
- Explain how sophisticated intracompany messaging capabilities, which disseminate information within the organization, have far outstripped their simplistic intercompany messaging counterparts.
- Describe the inherent problems in paper-based business documents, how companies have sought improvements for these problems in the past, and how electronic data interchange (EDI) and electronic commerce promise to be the solution for the future.

For many reasons, the current business environment is causing companies to rethink their traditional business procedures and practices.

Among these reasons are

- Intense competition here and abroad
- The *quality* movement, both in product and information
- The move toward more distributed processing within an organization
- The increased importance of rapid and accurate dissemination of information within and between business entities

Taking a look at each of these, we see that major US industries are looking for cost-cutting measures to make them competitive with foreign suppliers. Some are moving to offshore locations where labor rates can be as low as one-fifth of domestic rates. Others are merging to take advantage of economies of scale and to secure lower prices and more favorable delivery schedules by becoming major customers to their suppliers.

In today's market,

- Foreign products are eating into market share traditionally held by American products.
- Eroding profit margins are cutting into expected profits.
- Extremely inelastic demand for our products is causing large downward swings in demand in reaction to small increases in price.
- Wholesalers and distributors are experiencing severe competition, which forces them to work on very slim profit margins.
- American salaries continue to be much higher than those in foreign countries.

In light of the current economic environment, what are the paths open to our corporations that will enable them to improve productivity and customer service and increase net profit margin? How can corporations use their assets more efficiently, reduce labor and product costs, eliminate time lags, and provide value-added services to their trading partners?

Many of the guidelines that were espoused several years ago for a healthy and thriving business have been adjusted or even reversed in the economic business climate today. Years ago economists believed that large bank balances and sizable inventory levels were signs of a healthy and profitable business. Today, we know that the key to a profitable business is to have all your assets working for you.

The trend has been for:

Financial managers to maintain bank balances only large enough to cover current commitments while investing other cash assets in interest-bearing financial vehicles

Plant managers to store only enough parts inventory to satisfy current manufacturing requirements, with additional parts ordered on a specified delivery schedule as needed

Functional business managers to streamline business procedures, reserving staff for tasks requiring human judgment while automating the handling of rote activities

Administrators to take advantage of information available to them on a local microcomputer, developing reports of past and current activity that help them to forecast future trends in product demand

Marketing and customer service managers to seek ways of becoming more responsive to customer requests and to fill orders faster and more accurately

All the above and running the business "lean and mean" with a minimum of head count are the current popular criteria for running a profitable business.

JIT Manufacturing and Information

In response to these trends, customers and their vendors are entering into complex win-win business arrangements that offer benefits to both parties. For example, many manufacturing companies are taking their lead from the Japanese and instituting Just in Time (JIT) manufacturing schedules.

Traditionally, companies retained high levels of inventory, in part to protect against poor product quality, supplier delinquencies, and faulty order entry. They even planned manufacturing steps to build inventories of component parts prior to assembling the complete product. Viewing old films of automobile assembly lines, one would see hundreds of fender assemblies moving down the conveyor belt only to be stored until needed in another assembly operation. However, maintaining a large inventory requires excessive storage capacity, and excessive handling of stored product, and it ties up large quantities of capital.

The underlying philosophy of JIT manufacturing is for parts and components

- To arrive just in time for the manufacturing procedure in which they are used

- To be of 100 percent quality
- To be procured at the best possible price

Additionally, in a JIT environment, manufacturing steps are arranged such that each operation produces what is necessary to satisfy the demands of the subsequent one. In several industries today, with automotive the most noteworthy, the major original equipment manufacturers (OEMs) are placing increased pressure on their suppliers to support the JIT demands of product quality and delivery schedules. At the same time, these major customers are drastically reducing the number of suppliers from which they purchase product. The assumption is that those suppliers that can support JIT demands will get even more business and will have an even closer relationship than before—the ultimate competitive edge. In practice, customers are indeed buying more from those suppliers from whom they receive high-quality product at the right time and for the right price.

Even outside of the manufacturing arena the JIT concept is catching on. Known as quick response (QR) in the retail industry and efficient consumer response (ECR) in the grocery industry, the central theme is to reduce inventories by receiving product just in time to satisfy customer demand. As part of a QR or ECR program, vendors have begun to offer the highest level of customer service—supplying product to replenish stock as it is sold, with no need of a purchase request from the customer.

These cooperative programs are working to change the traditional customer-supplier relationship from one that has been traditionally somewhat adversarial, with each party trying to impose requirements on the other, to one that is based on cooperation and mutual respect. The "I've got to have it by..." and "You can't possibly get it until..." syndrome that was evident in trading relationships in the past was never conducive to cooperative intercompany business relationships.

However, in order to effect the timely and accurate JIT shipments of goods, transfers of business information must also be handled just in time. This means that manufacturers and other large customer organizations are requiring their suppliers to support technologies with which they are currently unfamiliar and for which they are ill-equipped. Sometimes, as these suppliers attempt to push the JIT philosophy down to their own vendors, they find that they have scant leverage with which to effect similar changes in business practices.

The Quality Movement

Also prevalent in business today is the trend toward *quality:* quality of product, measured by adherence to functional and testing specifications; quality of service, measured by a focus on customer needs and faster and more accurate response to inquiries and problems; quality of information, measured by its timeliness and accuracy as well as its accessibility to all those in need of it; and quality of process, measured by empowering the right people to do the job in the right manner at the right time. The overall trend in quality has greatly impacted traditional business practices and procedures, particularly in the provision of timely and accurate information.

In fact, not only have timeliness and quality of information become a major issue, quantity of information has increased immensely as well. Today, there is a move to share more information between trading partners than has heretofore been made available between companies doing business. For example, traditionally a customer organization would send an order to its vendor to effect a purchase. More recently that same company might precede the order with a forecast of its expected need for product over a specified time period and augment the order with sales or inventory-on-hand figures. The vendor might respond not only by shipping the goods and returning an invoice, but with an acknowledgment of the order and preshipment notification of what to expect in the impending shipment.

In the current paper-based and predominantly manual environment, the promise of more information implies more staff and considerably more time in which to process it. In addition, the accuracy of paper-based information is typically low since 1 in every 20 pieces of paper has a material error, such as incomplete or incorrect data, in it. Added to that, most companies experience a key-entry error rate of between 3 and 5 percent. So even when we receive correct business data, we may introduce errors during the manual processing and key-entry tasks. The cost to organizations of handling errors is astronomical. It is estimated that handling of an error condition costs 5 to 10 times that of handling correct information.

From a Centralized to a Distributed Environment

In conjunction with the need for more and more timely information, the shape of many organizations has changed from a predominantly centralized one, with most information access, processing, and maintenance done through one corporate data center, to a distributed arrangement where information is owned, processed, maintained, and accessed by the department or party that needs and uses it. This has occurred partially because the multitude of system enhancement requests have caused an unmanageable backlog of data processing projects in corporate information services departments and partly because of the availability of inexpensive and fast-running personal microcomputers (PCs) with huge memories and storage capacities. With the added ability to network several PCs through local area networks (LANs), banks of localized microcomputers are often acting as and taking the place of central computer installations.

However, along with the need for localized processing and access to information, there has arisen a need for the sharing of computer output between previously disparate computer applications that often reside on computers that do not easily "talk" to one another and for providing computer output to the business user in user-friendly online systems.

Intracompany versus Intercompany Messaging Capabilities

All this has caused business systems and data handling within an organization to rapidly become sophisticated, moving from their humble, manual processing beginnings to computerized processing, and from storage and handling of data in simple, sequentially organized piles and files to databases.

In fact, sophisticated companies today have responded to the need for dissemination of intracompany messages to various departmental business users, elimination of the rekeying of data, and maintenance of an accurate audit trail of intracompany activity by upgrading their files of business data to databases which can be accessed randomly using one or a combination of various access keys. Once housed in a keyed database, data are available to be used as input to more than one business application, with no key entry required.

In comparison, what then is the intercompany messaging environment? Even though intercompany transactions are the lifeblood of a company, their structure and handling have remained virtually static until fairly recently. Even those organizations that have sophisticated computer applications and internal database file structures still conduct intercompany business via standard business transactions, usually in the form of paper documents, telephone calls, or facsimiles. One or more of these typically accounts for the vast majority of intercompany messages covering all aspects of business; prepurchasing, purchasing and receiving, selling and distribution, and billing and paying.

Problems Inherent in Paper-based Business Documents

Companies that continue to trade using the traditional paper-based business documents still accept their lag time of up to several days while in transit via regular mail and the substantial amount of handling, key-entry, and processing time required at both sending and receiving sites. In addition, since their manually prepared business transactions are often received with inaccurate or incomplete information and are prone to errors being introduced during editing and key entry, these companies accept the added costs and time required to handle them. So, the handling of paper documents, being very labor intensive, is a particularly expensive medium in which to deal. Some see electronic trade as the answer to all these problems. However, this is not the first time that the business community has sought to improve the speed and accuracy and decrease the labor intensiveness of their procedures. There have been several interim solutions implemented by companies to alleviate one or more of the problems. For example, companies have increased the speed at which business documents reach them by

- Reducing in-transit mail time through the use of local post office boxes, lock boxes, and regional bank accounts
- Processing incoming transactions in disbursed locations, often local to trading partner locations.

They have increased the speed at which paper transactions are handled and processed by

- Developing on-line, menu-driven key-entry systems that edit data for completeness and correctness
- Developing proprietary computer systems which handle transactions at the trading partner site and route them directly to the company's site upon completion

Companies have tried to eliminate many of the inaccuracies associated with reading, interpreting, and key entry of incoming information by

- Developing preprinted forms to provide a common, recognizable format
- Developing on-line edits in key-entry operations to catch both the errors introduced during key entry and those received on the incoming business document

- Using product codes to completely describe a product and setting up a product code cross-reference file to convert trading partner codes received on incoming documents to the company's corresponding in-house codes
- Developing proprietary systems that support specific functions such as inventory inquiry or ordering in a user-friendly, on-line series of menus and preformatted screens

Finally, companies have tried to make the handling of transactions less labor-intensive by

- Developing computer applications to automate the tasks and decisions previously performed by people
- Passing processed data from one computer application to the next with no editing and key entry in between
- Automating reconciliation procedure such as that between the open purchase order, the incoming invoice, and receiving information

Each interim solution represents an improvement over the manual processing of paper-based documents. The logical endpoint would be to generate a data file out of a computer application and transmit it to the receiver in a format that, once received, could be fed directly into and automatically processed and acted upon by a receiving computer application. Unfortunately, the computer applications that process our business transactions were neither developed to generate nor accept EDI-formatted data. However, with EDI translation software in place to handle the EDI standard format, the second best is possible, that is, translating outgoing transactions generated by a computer application into the EDI standard format prior to transmission and translating incoming EDI transactions into a format acceptable to the receiver's computer applications as soon as they are received.

In fact, a growing number of companies (about 35,000 in the United States) have developed electronic intercompany trade arrangements with their major trading partners. Sometimes these were initiated as proprietary systems, supporting the internal business application of the initiating company. For example, Kmart, a major customer organization, offered, in the early 1980s, to make purchase orders available to its vendors in the predefined proprietary format generated by Kmart's purchasing application. This has been the typical pattern, that of customer organizations exerting pressure on vendors to satisfy their demands. Having the clout, they may choose to dictate how they want to conduct business and may even sweeten the pot by promising a closer trading relationship or a competitive edge to those suppliers that respond quickly to their requirements. Sometimes a major supplier exerts similar pressure on its customers by dictating how and when the purchase order must be delivered to them. In this case, they usually promise (and deliver) higher service levels and closer trading relationships for those that comply.

Additionally, some industry groups, through cooperation of their members, have developed electronic trade guidelines for companies in that industry, which identify the data needs of business transactions in electronic form.

Until recently, unless faced with a specific business reason for doing so, few companies implemented EDI. However, the EDI climate has changed. Because of the tremendous need to streamline intercompany business transactions and also as a result of very real benefits derived by those that have adopted electronic trade, many more companies and industries are currently investigating and defining their EDI needs. In some cases companies that are not yet able to support EDI are finding themselves at a marked competitive disadvantage and sometimes are unable to enter or compete in certain markets at all.

In fact, in some industries EDI is considered only the beginning. Today more and more companies are viewing EDI as step 1 of a three-step process culminating in the adoption of multimedia electronic commerce using various electronic capabilities such as EDI, electronic mail (E Mail), computer-aided design/manufacturing (CAD/CAM) drawings, facsimiles, voice-recognition computer applications, and access of machine-readable catalogs.

This trend is evidenced by the throngs of interested people attending EDI trade shows [typically over 500 at the annual Data Interchange Standards Association (DISA) trade show, over 350 at the annual EDI Coordinators' Conference, several hundred at the Uniform Communication Standard/Voluntary Interindustry Communications Standard (UCS/VICS) and American National Standards Institute (ANSI) X12 standards committee meetings, and at the many new EDI and electronic commerce conferences sprouting up each year].

The interest is high; however, the overwhelming majority of companies still support only traditional paper-based business transactions or are just beginning to dabble in EDI. Current EDI activity has hardly reached the tip-of-the-iceberg level, and electronic commerce is still only a vision for most. Many experienced EDI professionals believe that EDI is stalled today. They credit two reasons for this. One, lack of education on the business benefits of EDI, and, two, not understanding what it really takes to realize the benefits. So, while few people today are unfamiliar with the concept of EDI, most do not understand how it can really change and improve business.

Chapter 2 will discuss EDI in some detail, explaining how it works and the types of opportunities there are to use it throughout organizations.

In Chap. 3 you will get the chance to compare your own company's readiness to implement and use EDI against the parameters that have proven to be indicators of successful EDI implementations in other organizations.

Chapter 4 discusses the seven categories of business transactions and the opportunities retailers, manufacturers, wholesalers/distributors, and financial institutions have for using them.

Chapter 5 identifies various business issues to be considered when evaluating an EDI opportunity.

In Chap. 6 you learn who should be involved in your EDI program and what their roles are.

Chapter 7 follows with a step-by-step approach to both planning for and implementation of EDI.

Chapters 8, 9, and 10 spotlight various industries that have implemented EDI, focusing on those that have developed complex electronic transmissions of data, using several EDI transaction sets along with other multimedia electronic forms. These chapters are meant to describe the business environment of each industry and how each is being supported by electronic trade. In the process of describing the industry's use of EDI, the impact of automation and electronic trade on the buyers, suppliers, and wholesalers and distributors in the industry is identified.

Chapter 11 focuses on the various components of electronic commerce, how they are being used today, and prognoses for the future.

While not all business issues can be discussed in this book, business managers may notice similarities between the situations discussed here and those they encounter in their own businesses. These parallels may then be used to help them to formulate a plan for entry into the EDI arena and to anticipate trading partner and intracompany attitudes.

Summary

This opening chapter has sought to lay the framework for implementing EDI in the organization as a response to the company's need to remain competitive in the world market and to maintain market share and profit margins in the domestic market.

The number of companies showing an interest in EDI has been growing rapidly because of the mandate to thousands of automotive trading partners to do EDI to support JIT manufacturing schedules, to thousands of mass merchandiser vendors to do EDI to provide higher service levels and support QR systems, and because of the competitive economic environment in general.

The pressure is on. Your organization may have already incorporated some of the interim solutions to assist them to transact intercompany business faster, improve the accuracy of business documents, and decrease the labor intensiveness of conducting everyday business. Even so, EDI can dramatically improve your business picture.

As you read on, don't be fooled, as so many business people are, by the high-tech sound of EDI. While there are communication and computer hardware and software components of an EDI system, and while your information services department will be a major player in getting it up and running, EDI is predominantly a business initiative. In all successful EDI implementations, the lion's share of the time and effort is in the business area, and the business community is the owner of the program. As a functional business person, your main goal in evaluating and implementing EDI should be utilizing incoming EDI information throughout your organization in new and creative ways. Read on to learn how to get involved and what to do.

2 Opportunities for EDI

Objectives

- Provide a working definition of EDI.
- Provide a common knowledge base of EDI by giving an overview of electronic trade between a buyer and a seller organization.
- Demonstrate the major benefits of EDI throughout the organization.
- Describe four types of business environments that are particularly suitable for attaining benefits from EDI.
- Discuss five characteristics that make business messages suitable for EDI.
- Trace the growth of EDI through the use of proprietary company standards, industry-specific standards, generic cross-industry standards, and finally to an internationally accepted standard.
- Describe various components of electronic commerce.

In the never-ending quest for the competitive edge, increased market share, and net profits, EDI is viewed as a panacea for all ills. It is certainly not that. Electronic data interchange is not a cure-all for insufficient computer systems or inefficient business procedures. However, when a company supports electronic trade between itself and its trading partners, it earns the opportunity to reduce the flow of paper with the inherent error rate, substantially shorten the sales-purchase-pay cycle, and improve overall efficiency through more effective use of business data.

Understanding What EDI Is (and Isn't)

Electronic data interchange is the *intercompany computer-to-computer* communication of *standard business transactions* in a *standard format* that permits the receiver to perform the intended transaction (see Fig. 2-1).

Intercompany

Intercompany refers to the electronic transmission of data between companies. In order for successful transmission and receipt of data, both trading partners must have the same communications capability. However, just as there is little to no standardization in the way companies transact business, there is little standardization in the way companies are set up to transmit and receive data. The net result is that most companies doing EDI employ the services of an EDI third-party service provider or value-added network (VAN) to act as a communications intermediary between themselves and their EDI partners. Value-added networks typically support a variety of communications protocols (software that controls a communications session) and line speeds. By employing a VAN then, each EDI trading partner need only support one communications configuration.

Another intercompany aspect of EDI is that it typically requires a great deal of coordination to set up, test, and support electronic trade between an organization and its EDI trading partners.

Computer to Computer

Technically speaking one can accomplish EDI by communicating data from computer to computer with human intervention at each end, the sender key-entering data and the receiver printing it. However, this type of stand-alone EDI application, not drawing data from a computer application at the sending site and not feeding a computer application

1. Intercompany
2. Computer-to-computer
3. Standard business transactions
4. Standard format

Figure 2-1. The four components of EDI.

at the receiving site, provides very little benefit other than beating mail float. Companies that support this limited type of EDI are, in fact, supporting a very expensive facsimile (fax) capability.

The goal in EDI surpasses simply transmitting data between companies. It is to provide the link between sender and receiver business applications with no human intervention at the receiving end. The typical paper-based business uses human intervention at both ends. The sender generates a paper business document through a combination of key entry and computer processing. The document is mailed to a receiver, who manually edits and key-enters the same data and then processes them in its computer application.

In the EDI environment the sending side remains pretty much the same. The generation of an EDI business transaction is augmented by manual key entry. However, instead of by mail, delivery to the receiver is accomplished via electronic transmission. The receiver passes the data to the receiving computer application for processing. See Fig. 2-2 which illustrates the flow of information when sending a purchase order via both paper and electronic means.

In EDI the responsibility for sending a complete and accurate business transaction rests squarely on the shoulders of the sending party and for reading, interpreting, and using the data on that of the receiver.

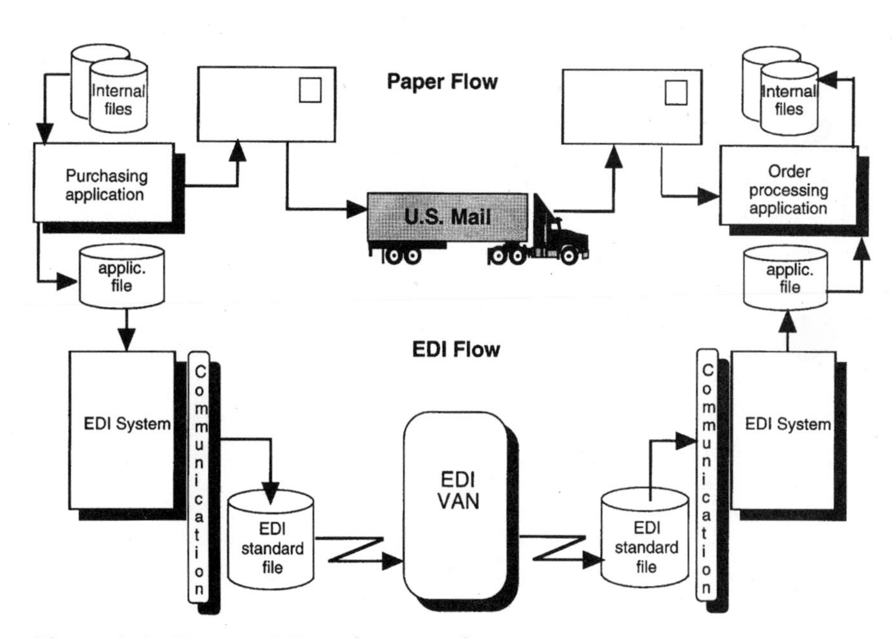

Figure 2-2. Paper and EDI information flow.

Unlike the paper environment, where documents are somewhat free-form in nature and often require human intervention to complete or correct prior to processing, machine-readable EDI transactions must be able to be read and interpreted correctly by a computer application. If they contain any ambiguities or errors, they are flagged as *exceptions*, rejected by the application, and referred to a person for correction. As you can imagine, this eliminates many of the benefits of EDI and reintroduces time lags and internally generated errors.

Standard Business Transactions

It must be made clear that EDI does *not* refer to the transmission of electronic mail or other free-form messages. The data referred to here are intended to be processed by a computer program, not a human being. Electronic data interchange transactions often replace printed business forms. They are designed to permit the receiver to perform a standard business transaction such as to process an order or to bill a customer. However, they are in machine-readable form, not in human-readable print format.

The goal in EDI is to send just enough information, in as abbreviated and codified a way as possible, to transact day-to-day business. This eliminates narrative information such as descriptions and free-form messages typically found on paper documents. However, to the extent that the narrative message conveys information that is needed to successfully process the transaction, senders and receivers must agree on codified ways of representing that same information in their EDI transactions. Because of this, there is a large amount of discussion and cooperation needed between trading partners to agree on the business information that will be included in each transaction and where it will be placed. This leads us to the fourth component of the definition of EDI, the standard format in which the data are transmitted.

Standard Format

In order for incoming EDI data to be recognized by the receiving computer application, the transmitted transactions must be in a predefined format. Because of the similarity in data requirements between companies, industries, and even countries to transact business, it has been possible to develop standards. Some standards have been designed specifically for application in one industry; one has cross-industry application; and one has international application. Each has been designed with a predetermined location for each data element, or field, needed to transact business in the company, industry, or country to which it pertains. And most have been designed with economy of space in mind, making it the perfect medium in which to transmit data. The standard that any one company supports for a specific business transaction is dependent on the company, industry, and country with which it trades. Examples of company standards are Kmart and Boise-Cascade. Examples of industry-specific standards are the Uniform Communications Standard (UCS) for the grocery industry and NWDA/ORDERNET for the pharmaceutical industry.

Companies in industries that have developed industry-specific standards need only adhere to the format, syntax, and usage rules of that standard to begin trading electronically with any of their industry trading partners.

With the widespread acceptance of the ANSI X12 standard to support cross-industry business transactions, most industries new to the EDI arena are choosing to support it instead of reinventing the wheel with their own standard. Nevertheless, while placement of data is defined in the standard, guidelines for usage and interpretation of data fields are being developed by individual industry groups. Companies in such industries, including electrical, chemical, metals, paper, office products, electronics, petroleum, and apparel, must adhere to the ANSI X12 format and syntax rules but may follow their own industry's guidelines for usage. When adhering to industry-specific guidelines, companies are actually using only a subset of the fields allowable in the X12 standard.

Companies that are in industries that have not developed industry-specific guidelines for usage but wish to support the ANSI X12 standard, must adhere to the ANSI X12 format and syntax rules and develop guidelines for usage directly with their trading partners. Organizations that trade with companies in several industries will sometimes be required to support multiple standards, or at the least multiple guidelines. Again, this will require a great deal of cooperation and coordination to ensure that sending trading partners know what they must transmit and receiving partners know what they will be receiving and where it will be located within the standard.

How Does EDI Work?

The concept is simple; EDI is the transmission of machine-readable data between trading partners' computers. Figure 2-3 illustrates the interchange of business transactions in electronic format. It shows two trad-

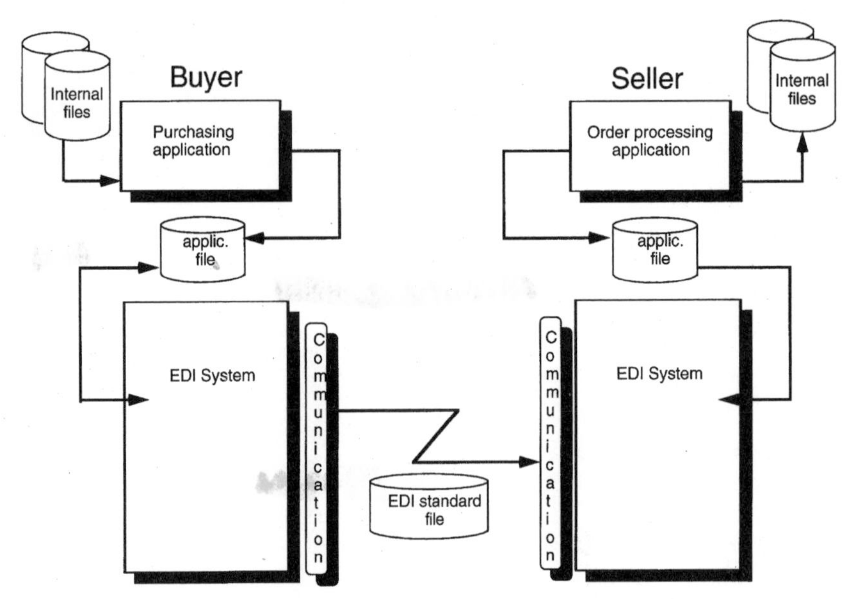

Figure 2-3. Buyer and seller EDI flow.

ing partners, a buyer and a seller, and the EDI data stream transmitted from the buyer's computer to the seller's. Let's suppose that the data stream is a group of purchase orders.

The buying company generates the purchase order transactions in its purchasing application just as it did in the paper environment. However, instead of printing the traditional paper document, it passes virtually the same information through the EDI system (which will be discussed in greater detail below), which generates the machine-readable EDI standard. This standard data stream is then transmitted to the seller's site. There it is passed through the seller's EDI system which maps the standard fields into the simple file needed by the receiving computer application, edits and verifies the incoming information, and then passes it to the receiving order entry application for processing. The order entry application processes it just as it would any incoming purchase order.

Important Aspects

Ideally there is a gateway computer application in place at each end of the trading pair. As shown in Fig. 2-4, *gateway* is used here in a special

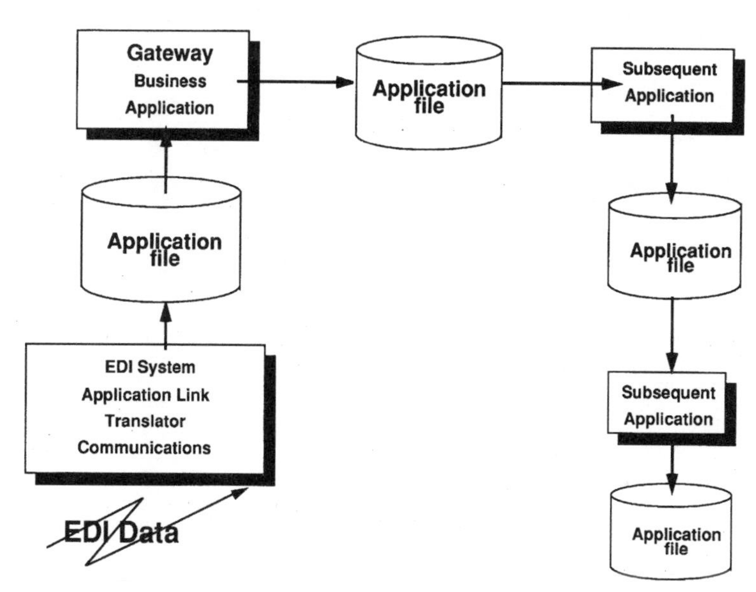

Figure 2-4. Gateway and subsequent business applications.

way. It refers to the last application a transaction passes through before being transmitted, such as the purchasing application, or the first application that the data reach after being received, such as the order entry application. This evokes the image of an electronic gateway through which data enter and exit a company.

Because existing applications are traditionally based on key entry of input and printing of output, they are rarely designed to accept and generate files and never to accept or generate the EDI standard format. Therefore, both the sender and receiver of EDI data must have computer programs in their EDI system that act as a bridge between their application and the EDI data stream. Typically there are two separate programs in the bridge. One is often referred to as the *application link*. As you follow this discussion in Fig. 2-5, you will see it on the sending side directly after the purchasing application. Here, its main function is to collect, from key entry and internal files, all of the information needed to generate the EDI transaction. From this information it generates a simple fixed-length file, so-called because all its data fields have predefined lengths instead of the compressed, variable-length fields of the EDI standard.

Still on the sending side, this fixed-length file is passed as input to the second of the bridge programs, the *EDI translator*. The EDI translator has two main functions. One, it uses the information from the fixed-

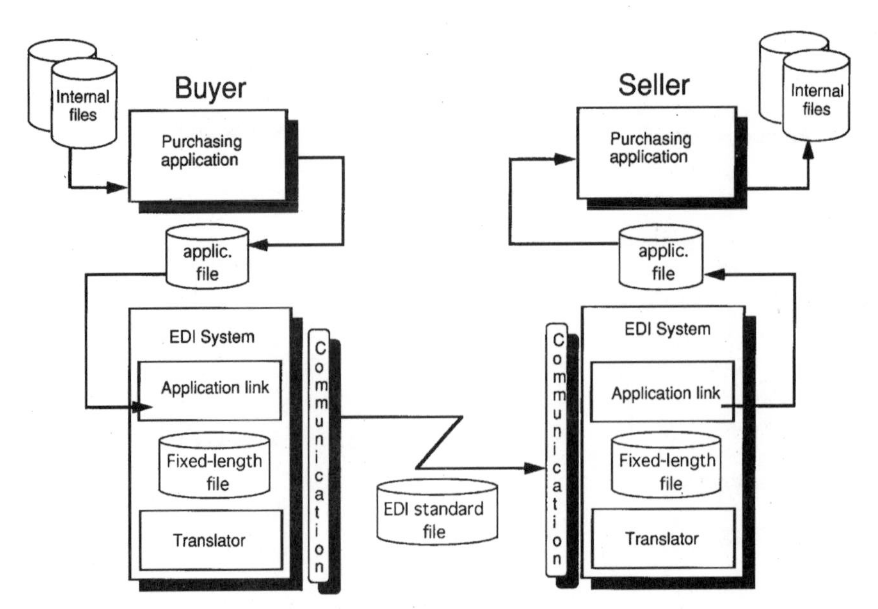

Figure 2-5. Bridging the gap between the business application and the EDI standard.

length file to generate the EDI standard file. Two, it assures that all the standard syntax rules have been met before approving the standard file for transmission.

Because the standard format represents an efficient communication medium with all the functionality needed to perform business transactions, EDI-active companies have agreed to generate their transmission files according to standard syntax and usage rules and to accept them in this format on the receiving side.

The EDI standard file is then transmitted via telephone lines, either directly from the sender to receiver or via a communications intermediary company called an EDI third-party service provider or VAN. In the typical EDI implementation both sender and receiver employ the services of a VAN because it eliminates the need for them to support different communications configurations with their various EDI trading partners and because it greatly reduces their internal support requirements.

As you can imagine, there are several VANs that provide EDI services. Sometimes, each trading partner employs a different one. In that case, the EDI data flow from the sender to the sender's VAN, on to the receiver's VAN, and then finally to the receiver. Most of the large EDI

VANs interconnect regularly with each other, so interconnects are common practice and cause no technological issue.

Still in Fig. 2-5, the receiving side, the data are received in the EDI standard format. Before being passed to the order entry application, data pass through two computer programs that represent the bridge to the receiver's application. First, the EDI translator performs two main functions. One, it verifies that the incoming data stream is complete and adheres to standard syntax rules. Two, it "maps" or moves the information in each data field from its location in the standard to its location in the fixed-length file.

On the receiving side, the fixed-length file is passed as input to the application link program. The application link, here, has two main functions. One, it makes available to the computer application a full transaction, in this case one purchase order at a time. In contrast, in the paper document and key-entry scenario, the application receives one key-entered field at a time which it verifies for correctness and completeness and reports errors back to the user via the terminal screen. Using EDI, the user interface is eliminated. So, the application must act on all the data fields consecutively, accumulating errors for a report that is passed back to a person for correction after the fact.

The receiving application link's second main function is to perform the editing and validation functions that were previously performed by people. For example, the application link would take over the task of verifying that all required pieces of information have been received, that internal product codes are substituted for customer product codes received on incoming purchase orders, and that incoming transactions are routed to the correct department or division. See Fig. 2-6 for a listing of the main functions of sending and receiving EDI system bridge programs.

Many companies eliminate this editing function and continue to keep people in the loop to perform the manual tasks that they were previously responsible for. While eliminating development of part of the application link is certainly a time and money saver, it also eliminates most of the benefits of EDI and greatly reduces the potential for savings. We highly recommend that you take this opportunity to rethink and automate your manual procedures as you implement EDI. However, be prepared for the fact that 90 percent of your time and money will be spent here on the receiving application link.

Following along in Fig. 2-5, the data finally arrive at the application. Here they are processed just as they were previously in the paper environment. As a matter of fact, the application remains virtually unchanged as it is almost completely buffered from EDI by the translator and application link.

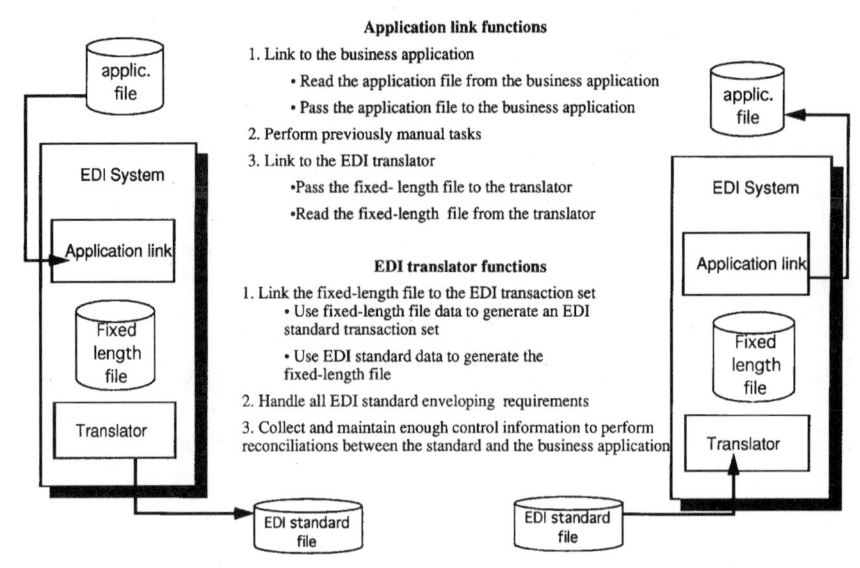

Figure 2-6. Functions of the bridge programs.

A Few General Notes

In the typical implementation of EDI, the application link is written in-house. This is the case because most applications were either written or customized in-house, and the application link output must be used directly by the application program. Also typical is the purchase of an EDI translator package. This is because the translator is a complex, difficult to write piece of code that interacts directly with the EDI standard. Since there are already many excellent, cost-effective EDI translators on the market for all sizes and almost all brands of computers, most companies choose to purchase one instead of embarking on the long and arduous road of developing their own.

Today several software developers are offering EDI front or back ends to their application software packages. On the sending side, these often preclude the need for application link and translation software by interpreting data fields directly from the application and mapping them between the fixed-length file format and the EDI standard file format. On the receiving side, these ensure that all the data elements needed by the application are passed correctly from the EDI data stream. However, they do not include the program logic needed to duplicate the manual editing and validating performed on incoming paper documents. Therefore, an application link will still need to be developed in-house.

Benefits of EDI

There are two types of benefits available from EDI. One is called *direct* because it happens as a direct result of implementing EDI. So, for example, a company implementing EDI would immediately realize the savings from needing fewer key-entry clerks and eliminating the need for envelopes and stamps. While these benefits are certainly real and realized in the short term, they usually do not represent savings large enough to justify major up-front development work and implementation of an aggressive EDI program.

The second type is called *indirect*. The indirect benefits of EDI are potentially much larger than the direct. However, they are realized, one, in the longer term and, two, as a result of a combination of initiatives comprised of both the implementation of EDI and the reengineering of business procedures and system processes. So the company that realizes this second type of benefit uses EDI merely as a conduit through which is passed timely and accurate information. However, it counts on the reworking of its computer programs, elimination of manual tasks and decision making, and drastic reduction in internally generated errors to realize major savings.

In general, when a company implements EDI, it is expecting

- Reduced costs
- Increased speed in information exchange and processing
- Shrinking of the order-receipt-pay business cycle
- Improved trading partner relationships
- Improved intracompany flow of information

Let's take a closer look at each of these expectations to see if they are, in fact, available from implementing EDI and, if so, what they require to be realized.

Reduced Costs Associated with Handling of Business Transactions

With EDI, you eliminate key entry of data and the errors introduced during that activity. You may also eliminate manual tasks such as sorting, matching, filing, reconciling, and mailing. Some of these are direct. For example, you reduce your key-entry requirements immediately upon doing EDI. However, most are indirect. You cannot eliminate a matching task unless you automate the reconciliation with a computer program that performs the matching, using the same logic and information sources as your staff does today. You can certainly reduce costs using EDI, but to realize substantial savings you must reengineer as well.

Reduced Costs for Material and Services to Support Paper Transactions

Upon implementing EDI you will decrease costs for paper, envelopes, and mailing materials as well as for telephone and courier services used to support intercompany business messages. Additionally, you will free up storage space for paper and material supplies and, eventually, for filing of paper transactions. These are direct benefits. As such, they are short-term gains but are usually not large enough on their own to justify a sizable investment in EDI.

Increased Speed

By eliminating use of mail service and by decreasing the time needed to process a business transaction, most companies see a definite improvement in the speed of information exchange when they implement EDI. However, increased speed is not a given for several reasons.

1. Unless a company manages its EDI transmissions and receipts efficiently, it may very well build in the same lag time already present in standard mail service. Many companies transmit and receive EDI data through a VAN. While this is not necessarily a problem, if they transmit irregularly or infrequently, they will fail to make their business transactions available to their trading partners in a timely manner. Likewise, if they pick up transactions infrequently from their VAN, they will fail to process them in a timely manner.

2. If a company does not automate all the manual tasks associated with incoming EDI transactions, the time saved during transmission is negated by the time still needed to perform manual editing and handling.

3. Typically, EDI transactions are more correct and complete than paper documents because they are generated by a computer program and undergo extensive editing by the EDI translator before transmission. However, some errors will still be transmitted because the sender's EDI translator looks only for syntax errors, not content errors. So, for example, if the sending company has the wrong product code stored in its internal files, it will use that wrong code when ordering product. If the receiving company does not plan for the speedy correction of exceptions found in incoming EDI transactions and the reintro-

duction of corrected data into the processing stream, the increased speed gained during transmission will be more than offset by the decreased efficiency resulting from handling errors.

Shrinking of the Order-Receipt-Pay Business Cycle

All things being equal, if an EDI order transaction is received sooner and is more likely to be correct and complete, it should allow product to be picked sooner, the order to be shipped sooner, and the product and invoice to be received sooner. In turn, this should lead to authorization to pay sooner and to payment received sooner. Do not misunderstand; payment is still dictated by payment terms. However, when companies receive the electronic invoice days sooner than they would have received the paper one, they are often able to reconcile their incoming invoices, authorize payment, and pay within the discount period. This helps both trading partners; payers qualify for the discount, and payees receive payment sooner than they would have previously.

This in turn leads to a secondary benefit, that of *reducing inventory levels.* As a result of reducing the number of days between placing an order and receiving the ordered product, companies can safely reduce their inventory levels. In addition, based on improving the speed and accuracy of information, companies can reduce their safety stock as well. A natural result is the placement of smaller and more frequent orders. Many EDI-active companies place daily or weekly orders instead of the monthly ones they placed prior to using EDI.

Improved Trading Partner Relations

Unlike other computerizations, EDI is not characterized by more distant and less personal relationships. While there are cases of one EDI partner unilaterally requiring EDI and specifying use of the standards, procedures, etc., in most cases partners have cooperated on how and when EDI will be set up and have agreed to the various applications that will be implemented. In fact, the most commonly heard benefit related by EDI-active companies is that there is a marked growth in the amount of human interface, information sharing, and cooperation between themselves and their trading partners.

One reason for this is that to automate transmission, interpretation, and processing of transmissions, a good deal of discussion and cooperation is required during the planning, implementation, and ongoing stages. With each party counting on the other to either provide accurate data or to process and act on data received, it is important that trading partners make agreements in the following areas prior to implementing electronic trade.

- Defining business procedures for the electronic environment
- Defining information requirements and specifying how the standard will be used
- Agreeing on common communications methods, line speeds, and *operational windows* (period of time during which data can be sent and received)
- Setting up testing schedules and agreeing on criteria for moving from test to production mode

Once EDI is implemented, trading partners must monitor its correctness and effectiveness, and tailor systems to improve both. In addition, each partner must continue to track EDI activity to assure that data are transmitted and received in total and on time at all communications points.

Improved Intracompany Flow of Data

As we have discussed thus far, the receipt of more accurate and complete business transactions in electronic form improves the flow of information and processing in the first application encountered. For example, the receipt of EDI purchase orders improves the processing speed and accuracy of the order entry application, and the receipt of electronic invoices streamlines the invoice verification and reconciliation process.

However, every EDI receiver gains the additional opportunity of gleaning information from incoming EDI transactions to use in subsequent applications and of developing and maintaining a complete audit trail of EDI activity. For example, incoming purchase order information can be used in an internal order to the warehouse from which it develops picking instructions. It can also be passed as input to a manufacturing requirements planning (MRP) system which analyzes the material requirements needed to fill the order and sometimes develops vendor purchase orders to satisfy those needs. In addition, purchase order information can be used to examine buying trends and, indirectly, to develop selling strategies.

An interesting statistic is that 70 percent of all the information ever shared between companies is found in the purchase order. So, receipt of an accurate and timely EDI purchase order can go a long way to populating many applications with the information you need.

It is just these additional uses for information, over and above the obvious use of the incoming EDI transaction, that provides the greatest long-term benefit of EDI. Nevertheless, most companies continue to use EDI transactions in very limited ways, mimicking their paper counterparts.

One other improvement in intracompany data flow deserves mention here. With the transmission and receipt of electronic transactions in EDI standard form, companies can automatically pull off control information from which they can develop a database of EDI activity. This information can be stored in such a way that it can be accessed randomly, through use of predefined access keys, by all those needing to use it. The first such user that comes to mind is the internal audit group. The ability to randomly access an audit trail of electronic documents, as needed, provides an audit capability unrivaled in the paper world. But others would also benefit from the same information. Take, for example, the incoming purchase order. The sales person would love to know that the customer order has been received. Customer service could use the incoming order to answer customer inquiries. As a matter of fact, by disseminating purchase order information to various departments within your organization, possibly in an on-line inquiry and reporting system, you provide the wherewithal to analyze existing business, develop forecasts for future business opportunities, evaluate costs and potential savings, and provide currently impossible services for your own personnel as well as for trading partners.

With the introduction of additional EDI transactions such as shipping documents and invoices, companies can more accurately prepare for receipt of goods, schedule manufacturing to meet demand, manage their cash, and pay their bills.

The possibilities are limitless. Within your own organizations, you can discover additional uses for incoming EDI information by asking the questions:

What other information would I like to get electronically?

What else would I do with the information?

Business Situations that Offer Excellent EDI Opportunities

There are four types of situations that are particularly suitable for conversion to EDI. Each is listed in Fig. 2-7 and is described below. However, just to be sure that you don't get the impression that converting a paper business document to electronic form automatically provides a solution, each section ends with the minimum you will need to do to your internal systems and procedures to realize any benefits.

1. Paper intensive
2. People intensive
3. Requires rapid information processing and delivery of goods
4. Processing bottlenecks

Figure 2-7. Four business situations suitable for conversion to EDI.

1. Situations that Are Paper-Intensive

Paper-intensive business processes tend to have certain characteristics in common. For one, they are typically slow. Paper takes a substantial amount of time to arrive and to be distributed. Even when a fax is used to speed up the process, it typically arrives at a central fax facility and still has to make its way to the intended receiver.

Second, paper documents are associated with a substantial number of errors because they are particularly prone to containing either incorrect or incomplete information. In fact, 1 in every 20 pieces of paper has been found to have a material error in it. When you consider the number of paper documents you send and receive, it all adds up to an alarming number of errors being handled daily by your staff or your trading partner's staff. And, when you consider that it typically costs five to ten times as much to handle an error condition as it costs to process a correct one, it all adds up to enormous error-based costs being borne as the cost of doing business.

Third, they represent a tremendous amount of duplication of effort because copies of paper documents tend to get distributed and used as separate entities by multiple departments and users within an organization. So, the accessing and handling of information as well as the filing and maintaining of paper documents is duplicated throughout the company. Even when we enter the information received on a paper document into a computer application, such as order entry, we rarely make the effort to share the now computerized information with other computer applications or with other users. Paper-based reconciliations tend to be the most paper-, people-, and error-intensive procedures within the organization, as exemplified by the company that performs a manual three-way reconciliation between outgoing paper purchase orders, incoming paper invoices, and shipment confirmation documents from its receiving department.

Finally, paper-intensive situations are particularly wasteful of resources because business is conducted through the use of turnaround documents. An incoming purchase order provides over 90 percent of the information for the outgoing invoice and shipping documents. In the paper environment, each functional business area re-creates the wheel. The sales area processes the purchase order, then accounts receivable uses a copy of the order to develop the invoice, and shipping uses another copy to generate the shipper. So, each department will individually perform data entry to get the same information into their various computer systems. With the time and resources used for these tasks along with the number of errors injected into the information at each turn, the average cost for processing an incoming paper document is reported to be $50 to $70. Some companies have even reported the cost to be well over $100.

Through the use of EDI, incoming transactions are received in electronic, machine-readable form. They arrive faster and can be distributed automatically. They are typically more accurate than their paper counterparts because they are generated by a computer application and because they are verified for completeness and adherence to standard syntax rules prior to transmission. The EDI transactions provide a database of timely and accurate business information that can be used to gain all the additional benefits associated with their processing and sharing. In order to actually automate the processing of EDI information in the first and subsequent computer applications, you must upgrade your computer systems to perform previously manual tasks and you must provide the link between one computer system and the next.

2. Situations that Are People-Intensive

People-intensive business situations are particularly good opportunities for conversion to EDI. Since salaries are extremely high in comparison to computer resources, companies can typically substantially lower their costs by automating manual procedures. Most often, in people-intensive business situations, people are being used as "paper pushers," performing somewhat rote tasks such as handling, reading, validating, and correcting information on paper documents. By eliminating the paper document, while providing access to the same information from a computer-readable and processible file, a company positions itself to automate the paper-pushing tasks. Take, for example, an organization that uses a staff of clerks to correctly route incoming paper documents to the appropriate division, department, and individual receiver; to manually edit and validate trading partner information; and to cross-reference specific fields throughout the document. These are just the types of tasks that they can easily be automated. The requirement is that they develop a program that can handle the tasks using the same logic and having access to the same information as the people previously did. Of course, when errors are encountered or complex decisions need to be made, people can easily be brought back into the information loop and given access to the information they need.

Remember, getting your business people out of the paper loop does not mean removing them from the information loop. Today business people conduct business using information that they find on paper business documents. When we eliminate the paper, we must provide them with access to the information that they need. This sounds like a small point; however, believe it or not, most of the time, effort, and dollars in your implementation will be spent right here, developing ways to get your business people back into the information loop in a user-friendly and useful way. More about this later.

3. Situations that Require Rapid Information Processing and Speedy Delivery of Goods

Here, the electronic environment is a virtual necessity. Even in the most elementary, bare bones EDI implementation, electronically transmitted transactions are received in a much more timely manner than their paper counterparts. Counting on increased speed, many organizations have developed new business procedures that require very fast turnaround of business documents to support equally fast turnaround of product.

An example is the move toward JIT manufacturing schedules that require product to be delivered just in time to be included in the next manufacturing process. Just as customers that support JIT manufacturing schedules are requiring very fast product turnaround from their vendors, they are requiring even faster information turnaround so that they are informed that the ordered product will be arriving on time. Vendors are hard pressed to provide the JIT information needed via paper documents, so they are implementing EDI. This allows them to respond to pressure from customers, to accept orders electronically from them, and to provide their intention to ship via an electronic advanced shipping notice. Electronic data interchange is more than a convenience in this situation. In some cases, such as in the automotive industry, it is a necessity for doing business. Similar programs are being developed in various different industries, e.g., QR in the retail industry and ECR in the grocery industry. These will be discussed in more detail later on.

Often, manufacturing companies combine vendor management activities (reducing the number of vendors from whom they buy and developing closer ties and larger commitments to those remaining) with requests for implementing EDI. The none too subtle implication is that those vendors who are EDI-ready will not only keep the business but will get more than their current share of the pie.

The warning here is that just receiving electronic transmissions does not automatically allow you to handle or respond to them more quickly or more accurately. In order to speed up your internal procedures, you must develop the computer programs and streamlined procedures that will allow you to more efficiently process incoming transactions, act on them, and turn them around into outgoing ones.

4. Processing Bottlenecks

Often, it is not the time it takes to process a paper document that slows down our business procedures. It is the time it takes to get the time to process the paper document. While it may take only 10 minutes to key-enter an incoming order, this task is rarely completed in the 10 minutes after receipt of the document.

Recently there was a keynote speaker from a large insurance company who related this story. His company decided to perform an analysis of the tasks and time required to activate a new insurance policy, from initial application to live policy. They discovered that there were 26 discrete steps required and that these steps typically took 30 days to complete. When they investigated how much value was added during each step (in other words, how much work was actually being done), they discovered that in all 26 steps combined, there was only one-half hour of value added. What this pointed out was that there was just a minuscule amount of actual work time during the elapsed 30 days. Most of the time was taken in waiting for someone to get to it. If you can picture this scenario with the insurance application received in EDI form, you would surely see how much more quickly and accurately the data could be handled. The resulting rapid processing and activation of new policies in a fraction of the time currently required would provide a definite competitive advantage for this company. Surely, examples such as this exist in all our companies. The benefit of converting these applications to EDI is that they can be scheduled to run as soon as incoming EDI data are received. Here, you'll not only need to develop the program that performs the currently manual task but also to schedule regular retrieval of EDI transactions and to initiate the processing as soon as data are received.

Just as there are situations that are particularly applicable to EDI, there are others that are not. One such example is sporadic or one-time ordering situations. This is so because setting up an EDI relationship takes a good deal of time and effort and is only warranted for an ongoing business relationship.

Another is ordering of custom-designed or custom-built product. Typically, transactions dealing with customized product require manufacturing specifications, graphics, blueprints, and narrative comments accompanying the request for quote and ordering transactions. While there is no technological issue in transmitting both EDI and non-EDI graphics, blueprints, etc., either in the same transaction or separately, most companies still choose to handle custom orders in a people-oriented, nonautomated way.

A noticeable exception is the metals industry, represented by aluminum and steel, which have attempted to quantify and codify all parameters associated with manufacturing specifications of their products. They have defined many additional data fields for use in their industry. In addition, they have designed a series of segments that can be used in the EDI purchase order that relate these manufacturing specifications along with the more traditional purchase order information. As you can imagine, the computer system receiving such a purchase order would have to be extremely "smart" in order to interpret and act upon such complex incoming data.

Business Messages that Are Particularly Suitable for EDI

We've examined various business situations and discussed whether and how they lend themselves to automation and EDI. Now let's focus on business documents and other forms of business messages such as telephone calls, facsimiles, and catalog access with which we currently conduct business. In attempting to evaluate their conversion potential to an EDI format, we find that there are five parameters that are particularly meaningful. They are:

- *Frequency.* How often the document is sent and received
- *Volume of data.* How much data are included in each document
- *Time-criticality.* How critical it is that this document be sent, received, and handled quickly
- *Content criticality.* How critical it is that the information contained in the document be accurate as sent and received
- *Predictability.* How well defined the transaction is and how easily its contents can be codified and narrative, free-form comments eliminated

Figure 2-8 shows a sample transaction matrix that you can use to rate your business documents in each of these five categories. Place a rating

Frequency	Volume	Time criticality	Content Criticality	Predictability	EDI Rating

Figure 2-8. Transaction rating matrix.

from 1 to 10 in the box for each parameter and develop an overall EDI rating for the message. Based on the ratings you arrive at for the various documents and other messages that you regularly use to conduct business, prioritize their conversion to EDI. Following is a discussion of how to calculate the EDI rating for each document. You'll notice that this is far from a scientific approach, but is meant to help you to compare your business messages.

1. Frequency

In general, those documents that you rate high in frequency are your best candidates for EDI. As discussed later, a good deal of time and effort is required to prepare your business and system environments for an EDI transaction. When that transaction occurs frequently, the effort pays off in two ways. One, it greatly reduces your currently sizable manual effort. Two, it leads to noticeably decreased internal costs and increased accuracy. An excellent example of a frequently occurring document is the purchase order. Even better is the invoice. Often there are 2 to 3 times the number of invoices as there are purchase orders. Still other frequently occurring documents are inventory and order status inquiries and responses.

2. Volume of Data

Sometimes conversion of an infrequently occurring transaction provides a good opportunity as well. In order to obtain savings from conversion of such a transaction, that transaction must either contain very large amounts of data or represent an inordinately large manual effort. For example, you may want to automate use of a product or price catalog. While this document is received infrequently, it usually contains a very large volume of information and requires a tremendous effort to key-enter. By examining the time and people resources you currently expend in handling and using a catalog, along with the error rate you currently experience in misread and miskeyed information, you will be able to evaluate your potential savings from receiving it in a machine-readable form and eliminating all the associated people and error costs.

3. Time Criticality

When a document is particularly time-critical, it is usually a good candidate for EDI as well. However, as mentioned above, getting the document quickly is only part of the solution. The goal is to handle it quickly through streamlined internal procedures. For this you will need to design your systems so that incoming transactions are processed automatically and passed to your internal applications immediately upon receipt. Because this system work is certainly not inconsequential, many companies choose to forgo the savings of EDI and to continue to handle time-critical documents such as purchase order changes manually, through use of supporting telephone calls and facsimiles.

4. Content Criticality

Most documents are content-critical. However, in some the information is particularly difficult to enter or verify for correctness. Documents supporting international trade are excellent examples, where names, addresses, and other related pieces of information are unpredictable and often in an unfamiliar format.

5. Predictability

Finally, a note about predictability. Those documents whose form is well defined and whose content can be predicted and is easily codified are much better candidates than those that are free-form with narrative content. Since the requirement in EDI is to program the sending application to generate the outgoing EDI transaction, and to program the receiving computer application to read, interpret, and act upon the incoming EDI transaction, you must know in advance what information to expect and where it will be located.

To the extent that the information can be abbreviated and represented in codes and identifiers, rather than narratives and descriptions, you increase the productivity of your EDI system, reduce your transmission costs, and eliminate errors. For example, you could include your full name and address in an EDI transaction.

ABC Company, Incorporated
112 Main Street
Any City, Any State, 12345-6789

or, you could include instead your customer identifier (ID) 987654, most likely assigned by the vendor.

In the first case, the receiver compares the incoming name and address to its own customer file to verify that you are a customer. Depending on how the company has stored that information, an exact match may or may not be found in the file. Even an inconsequential difference such as *Inc.* instead of *Incorporated* would look like a nonmatch to a computer and would cause the incoming transaction to be handled as an *exception,* an error condition. In the second case, only a match on your customer ID would need to be found and then your name and address could be picked up directly from the file.

As another example, you could include a narrative product description when ordering a product or provide a product code that completely identifies the product. In the first case, the receiving application would need to read and recognize the description in order to understand and process the order. Any difference in spelling or abbreviating would require intervention of a person. In the second, only the agreed upon code would need to be found in the file.

The ability to codify information becomes even more important in international trade than it is domestically as it eliminates language barriers. In the above cases, sending the customer ID and the product code instead of the name and address and product description completely eliminates the need to share any language-oriented information.

Rating your business transactions as high or low based on these parameters will enable you to develop an EDI rating for each. Sometimes, a low value of one parameter is balanced by a high value in another and provides surprisingly high paybacks as EDI transactions. For example, an inventory inquiry is very low in data volume, requiring only product descriptor and quantity, but may be high in frequency, regularly preceding placement of an order. On the sending side, its low volume of data requirements would make implementation particularly easy. On the receiving side, minimal data requirements would make it easy to act upon and convert into the turnaround inventory advice transaction. Not only that, but upon receipt of a favorable inventory advice, this transaction can be further turned around into an EDI purchase order.

Another example of surprisingly high payoff in spite of transaction indicators is conversion to electronic format of manufacturer deal and promotion offers. Regularly, food manufacturers send to their customers paper documentation describing the wide variety of deals and promotions they are currently offering. In response to these offers, grocery chains employ staffs of employees who analyze them and try to make the most cost-effective buying decisions. Because of the complexity of the manufacturer deals, they often make good, but not the best, buying decisions. As a result of an analysis of deal-handling activities, some large chains computed their costs and estimated the savings they could realize by always making the most cost-effective buying decisions. One major chain estimated that they could save $600,000 to $700,000 per year from automating and optimizing the buying decisions they make.

As you can imagine, it was no easy task to develop an electronic format for deals and promotions, with all their associated parameters, such as qualifying product, quantity, dates, geographical location, display requirements, signage, and complementary product purchases. However, it turned out that developing the electronic format for deals and promotions was the easy part. In order to get any real payback, a company would need to automate its processing. Development of a new business application to automate handling of EDI deals and promotions would require a large dollar investment as well as a great deal of time and effort. When the costs were compared to the potential savings, the up-front investment was actually the most cost-effective approach.

The moral of the story is that a low EDI rating for a business transaction based on the parameters described above may be incorrect if you also identify the transaction as being particularly expensive to process. As in the above cases, it may be worth your while to bite the bullet with extensive and expensive system development to take advantage of potentially high savings.

The next step is then to prioritize your transactions by their ratings and plan to convert them to EDI starting with the highest priority transaction.

Growth of Electronic Trade

Today we think of EDI as something new. Of course, intercompany sharing of information has been around for as long as businesses have existed. The new parts are electronic transmission, which merely uses the latest available technique for sending data, and the machine-readable aspect. Electronic data interchange is just another attempt at speeding up intercompany business procedures.

It is interesting to look back at the first attempts at trading electronically, tracing the evolution thus far, and, from that, predicting the directions and activity we expect of EDI and electronic commerce in the future. Electronic data interchange was initially introduced as a medium for the electronic transmission of data by two types of companies, one a major supplier, the other a major customer. Each offered its trading partners an EDI system that supported its own proprietary company format. All did it to improve their own business situations, and all had the clout to influence their trading partners, as they represented a major portion of the trading partner's business. Let's look a little more closely to discover what each stood to gain from promoting its own proprietary system and how its trading partners fared.

Proprietary Standards

The *major supplier* offered a proprietary ordering (purchasing) system, as shown in Fig. 2-9, to its large customers; they often threw in hardware and software to handle the ordering function. Usually this company had a centralized order-processing system that handled all incoming orders.

A typical proprietary ordering system accepts key-entered information in an interactive system at the customer site and generates and transmits orders in the proprietary format required by the supplier's order entry system. However, sometimes the supplier's order entry system will also accept an order file directly from the customer. In this case, the supplier provides documentation explaining its file format requirements, and the customer is expected to transmit the appropriate file to the supplier's system.

Obviously, the customer deals with more than one major supplier. If each offers its own proprietary ordering system, the customer soon has a half-dozen microcomputers with as many different menu systems, product numbering schemes, order formats, and communications requirements. Soon the convenience of each becomes a tremendous has-

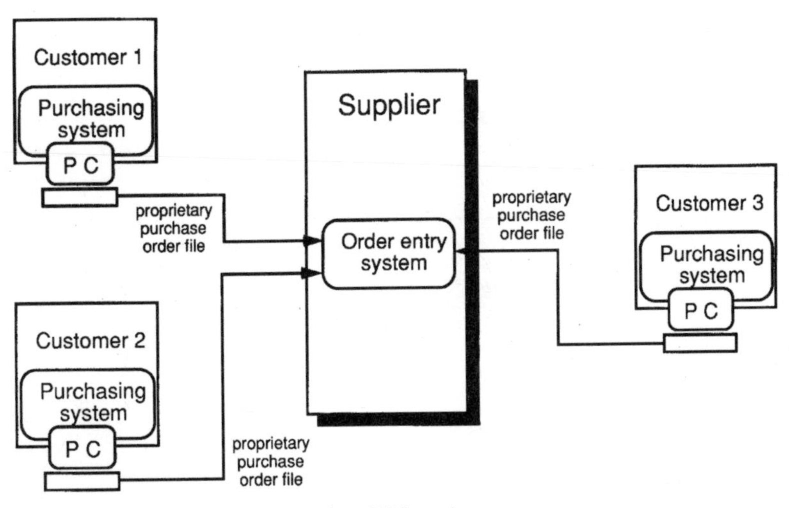

Figure 2-9. Proprietary supplier EDI system.

sle. And, while each of the systems directs order information to a supplier, none directs any data into the customer's inventory system or other internal applications. In order for customers to maintain data on what they have ordered, they have to turn around and key-enter the data into their own internal systems.

However, the good news is that in return for using the proprietary system the customer is promised better service with which it can gain an edge over its competitors. In fact, in many instances the customer does realize benefits from the use of such a proprietary system, sometimes through on-line inventory and order status inquiry capability and often through more timely deliveries.

On the supplier side, there are some very definite benefits from offering such a proprietary system. For one, the supplier differentiates itself by promoting a customer-service–oriented image and making itself more convenient to order from and to do business with in general. For another, the supplier substantially cuts its order-processing costs. For still another, it tends to tie in its customers through use of the system, which results in larger and/or more frequent orders than previously.

The situation is similar with the proprietary system, as shown in Fig. 2-10, offered by *major customers* to their suppliers. With a centralized purchasing function, such a customer is a force to be reckoned with. In this type of arrangement, suppliers are usually required to call in (sometimes at specified times) and pull back awaiting orders. Those orders are available in the proprietary format of the customer's purchasing application. In order for suppliers to process them, they must interpret

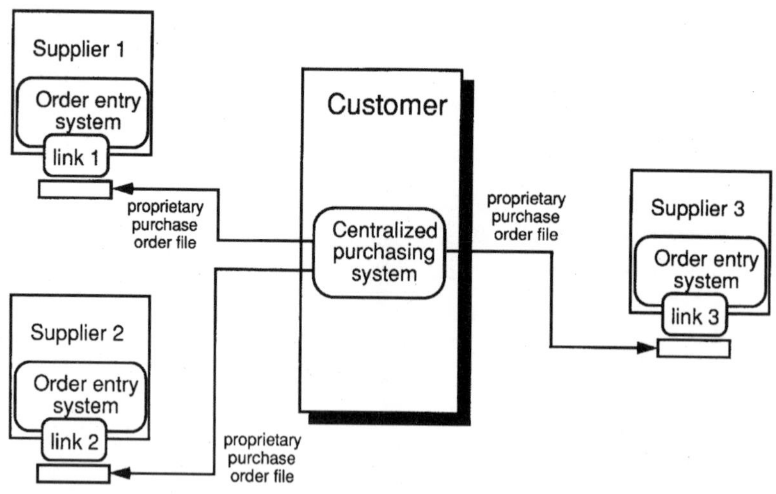

Figure 2-10. Proprietary customer EDI system.

the incoming data, reformat it or key-enter it into their own order entry application, and then process it just as they would any other order.

Since pressure to support the proprietary purchasing system is coming from a major customer and is often accompanied by threats or promises of giving favored treatment to the supplier, there is rarely a need to offer hardware or software to convince suppliers to use the system. In fact, experience shows that the customer often does increase the amount of business it does with suppliers who support the proprietary system. Some large customer organizations have even used the ability to support EDI as a criterion for selecting or keeping a supplier.

In order to support a proprietary format, suppliers must develop a front-end link to reformat the proprietary data-field placement into the internal format required by their own order entry applications. When communications also is handled by a proprietary format, suppliers must also purchase software and hardware that supports these communications requirements. All in all, support of proprietary requirements may represent a substantial amount of software development and support by the vendor and may be perceived as returning them fewer benefits than it does to the customer. Proprietary company systems, in general, offer greater benefits to the initiator of the system than they do to the trading partners.

Industry-Specific Standards

The next logical step in the growth of standards was to look for opportunities for more companies to support one standard. In trying to identify groups whose members conduct business in a very similar way, an industry was a logical unit. Consequently, an industry format is based on the homogeneous data requirements needed to handle business transactions within that industry. Different industries have developed different standards and methods of communications. Industry-specific standards are developed to support the business requirements of buyers and sellers in the industry. Usually an active trade organization sponsors or supports the effort. With broad industry participation in their development, transaction formats contain all the data fields needed by all companies in the industry to conduct business. Additionally, the requirements and method of communications are developed with the industry companies in mind, so they work for all sizes and levels of sophistication of participating industry members.

For example, the pharmaceutical industry, spearheaded by wholesalers and the National Wholesale Druggists Association (NWDA), developed a fixed-length standard format in the late 1970s for the purchase order and invoice transactions. At that time the NWDA endorsed one communications VAN through which all transactions would flow. Components of the communications session such as communication protocol and line speed were left to the discretion of the participants and the VAN. Later, this industry developed formats for additional transactions, most notably, some that never before existed in paper form. Chapter 9 spotlights the healthcare industry and describes its business conditions, competitive environment, and proprietary and industry EDI activities.

Another example, the grocery industry, led by the major grocery manufacturers, retailers, and brokers, and the Uniform Communications Council (UCC), developed in the early 1980s a variable-length standard for grocery industry purchase orders, invoices, and various other transactions. These same companies were also active in developing the Warehouse Information Network Standard (WINS) for public warehouse documents. In addition, they used some transportation documents which had already been developed by the Transportation Data Coordinating Committee (TDCC). The preferred communication method for direct transmissions between trading partners was bisynchronous 2780/3780 protocol at 2400 baud. The grocery industry is another that will be handled in depth, as part of Chap. 8 on the retail industry. It will be interesting to note how these two industries have progressed in EDI considering how different their early decisions and development were.

You can see the benefit of developing an industrywide EDI capability in Fig. 2-11. The full trading community of manufacturers, distributors

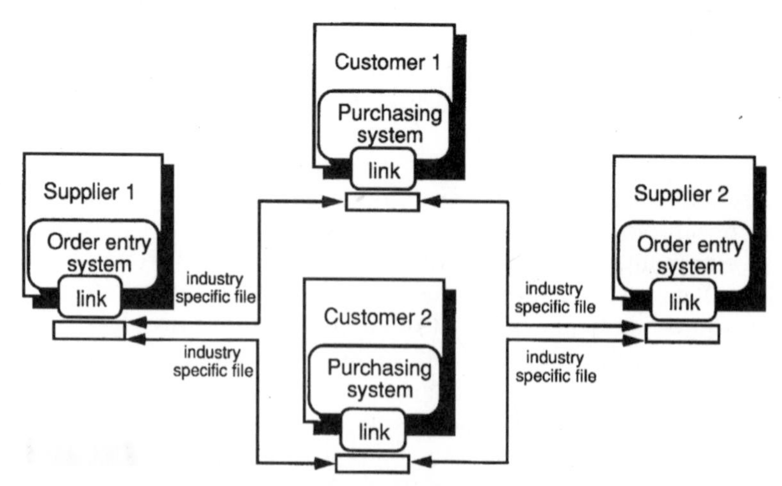

Figure 2-11. Industry-specifiec EDI system.

and wholesalers, and retailers can implement EDI using the same standard. Therefore, any industry member wishing to trade electronically with others in its industry can do so by developing one application link between its own business application and the prescribed standard cosupported by all its trading partners.

As benefits of EDI accrue in direct proportion to the amount of business performed electronically, industrywide EDI implementation provides potentially very high benefits.

Cross-Industry Standard

The next step in the logical progression from company standard to intraindustry (within one industry) standard was to an interindustry (across industries) standard. Since most companies deal with trading partners from many different industries, what was needed was a generic standard that would support multiindustry electronic trade. Just as the intraindustry standard is based on the homogeneity of business requirements within an industry, the interindustry standard is based on the mostly homogeneous business requirements across industry lines.

The American National Standards Institute (ANSI) became the sponsoring standards organization for the interindustry standard. The Accredited Standards Committee (ASC) X12 was given the mandate to develop variable-length data formats for standard business transactions that would have interindustry application. The ASC X12 committee, accredited in 1980, has members representing a multitude of buyer and seller companies from many different industries, including government. In addition, there are representatives of the various third-party providers of software, network services, consulting, and education in EDI.

Because an interindustry standard must support the functionality required by all industries, X12 EDI transactions tend to be more complex than any industry-specific or company proprietary standard. Additionally, this standard contains many more types of transactions than any one company or industry uses to conduct business. However, when companies need to trade with partners from several different industries, and when they need the extra functionality in their business transactions, they can use the generic X12 standard. What's more, use of this standard allows them to support EDI with all their partners by developing only one application link between their application and their trading partner base. This scenario is illustrated in Fig. 2-12.

The X12 standard has developed an extensive following over the last several years. For this reason, those industries that have not already developed an industry-specific standard are today supporting the X12

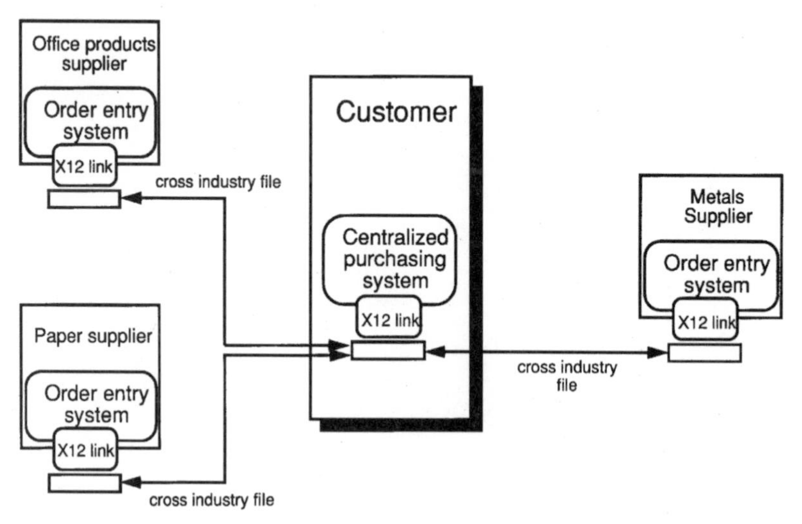

Figure 2-12. Cross-industry EDI system.

standard. In fact, just recently the TDCC (transportation industry) standard, the UCS (grocery industry) standard, and the WINS (warehouse) standard have all moved under the X12 organization's auspices. An interesting phenomenon has been the growth of industry guidelines for usage of the X12 standard. Because of the complexity of X12, several industries have sought to limit their exposure to the standard by specifying exactly which parts of the standard they will use to support business requirements in their industry. They have then developed industry guidelines for use of the standard. Using the industry guideline, member companies need only prepare to send and receive a subset of the data fields available in the full standard.

Some examples of industries and their trade organizations that support X12 and have developed industry guidelines for its use are given below. For the full names of these organizations, refer to the glossary at the back of this book.

Automotive	AIAG
Retail	NRMA, VICS
Chemical	CIDX
Electronics	EIDX
Electrical	EDX
Petroleum	PIDX
Metals	Aluminum Association, AISI

Paper	API
Office products	NOPA, WSA

International Standard

Finally, in the progression of EDI standards comes the international standard, supported by trading companies in many countries. For this purpose, EDI for administration, commerce, and transport (EDIFACT) was developed. Following is a list of some of the important events in the evolution of international EDI.

As far back as 1961, the United Nations Economic Commission for Europe (UN/ECE) Committee on Development of Trade established a working party on the simplification and standardization of external trade documents, which became the first instance of a development effort for standardization of international trade procedures and documentation.

In 1968, the Transportation Data Coordinating Committee (TDCC) formed with the support of the US Department of Defense to develop messaging standards for interorganizational communication of business transactions, became the first developers of intercompany EDI transactions in the United States. The standards resulting from this effort were and still are used by the rail, motor, ocean, and air carrier and shippers industries.

In 1979, the ANSI chartered the ASC X12 to develop a standard for the interindustry electronic interchange of business transactions. The syntax and structure of the resulting standards were based on those developed by TDCC but contained the functionality and data content to allow them to be used by all industries and organizations.

In 1983, major auto manufacturers from the United Kingdom, Italy, France, Belgium, Sweden, the Netherlands, and Germany formed a committee to develop EDI standards for the automotive industry in their respective countries. That committee became the ODETTE EDI Standards Committee the following year. The automotive industry still represents the largest group of EDI users in Europe.

In 1984, ANSI X12 and TDCC formed a joint electronic data interchange (JEDI) committee to consolidate the data element dictionaries and data segment directories of the two standards. At the completion of this task, TDCC became the official gatekeeper to assign new data element IDs, new data element code values, new segment IDs, and new transaction set IDs for both standards.

In 1985, after a year and a half of work, ODETTE with the assistance of the Automobile Industry Action Group (AIAG), its US counterpart, developed standard message formats for the European auto industry. Then, in November of 1985, North American and European EDI interests (from more than 20 countries) assembled to discuss the feasibility of establishing a joint syntax for an international standard format to be used in international trade. They met again in March of 1986 and then at least monthly for the next 18 months under the cochairmanship of Dennis McGinnis of North American Philips and R. J. Walkers, chief executive of the Simplification of International Trade Procedures (SITPRO) board. The result is an international EDI syntax called EDIFACT.

The EDIFACT syntax became a draft international EDI standard by unanimous vote. The EDIFACT data element dictionary would be identical to the *United Nations trade data element dictionary*, and standardized code lists would coincide with the International Standards Organization (ISO) code lists where they exist.

EDIFACT messages are under development on several fronts. The invoice was the first to be approved as a universal standard message (UNSM) in December of 1987. Today there are over 100 messages in various stages of development and over 80 already approved and being used. Messages have been developed for various areas of trade. The primary focus is on transportation. However, there are several that deal with finance as well.

EDIFACT is destined to gain strong backing from North American, Western European, Eastern European, and Pacific Rim business interests because it is in the unique position of having been developed prior to specific business practices becoming so entrenched that they could not be changed. The thing that makes EDIFACT particularly useful is its focus on business function instead of replacement of specific paper documents. With the focus on function, transaction sets are developed as a series of predefined segments containing the information strings required to transact a particular business transaction. Each party to the transaction will find in the segments all the information it needs to transact business.

A tremendous boon to international support of EDIFACT came when the US Customs Commissioner, William Von Raab, announced early in 1988 that the US Customs Service would support the EDIFACT standards developed by international business interests. Customs had previously been talking of developing a proprietary auto-broker interface which was being met with criticism as an inappropriate direction to take at the time. Following this lead, the United Kingdom and Australia announced their support of EDIFACT. With this kind of government support, companies considering implementing EDI to handle international trade can now be certain that there is a standard for message formats that is universally accepted, that that standard is EDIFACT, and that the business community will now begin to move rapidly toward an electronic environment.

EDIFACT has been described by Alice Rigdon, customer representative to the European community in Brussels, as "a snowball coming down from the top of the hill that got a big push from Customs Commissioner Von Raab." It is agreed that EDIFACT will be a fact of life for international trade communications.

How will generation and interpretation of EDIFACT transaction sets fit into the domestic EDI systems that corporations have established? While EDIFACT syntax is very similar to the ANSI X12 syntax, some differences do exist. Most mainframe and midrange computer translation software products on the market today can handle international as well as domestic standards. Homegrown software would have to be enhanced to handle new syntax situations. In addition, data requirements for international trade are more extensive than those for domestic transactions. The sender of an EDIFACT transaction would need to provide data elements required for the international message standard in order to generate a complete transaction; the receiver would need systems to interpret and act on the expanded set of data fields he or she will receive. While some additional application link programming might also be needed to handle EDIFACT standard messages, neither the syntax nor the data differences are extensive enough to cause major problems to an already existing EDI application.

Certainly the potential benefits of eliminating paper in international trade would far outweigh the extra development costs required to support the EDIFACT standard. Figure 2-13 illustrates how international companies would use EDIFACT to trade electronically.

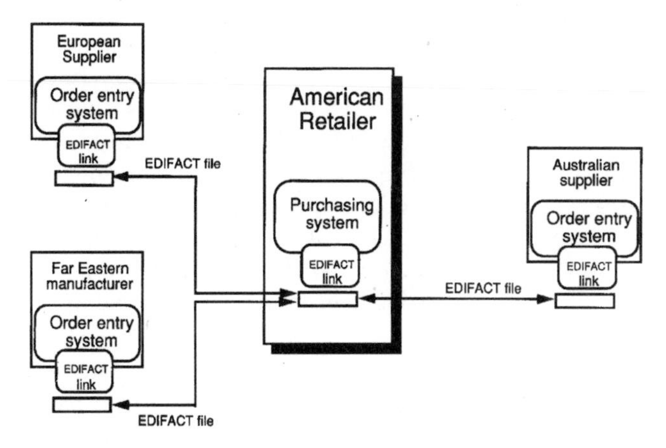

Figure 2-13. International EDI system.

Other Types of Electronic Messaging

It is important to note that messaging or sharing of information occurs both intracompany and intercompany, with EDI being the intercompany, machine-readable component only. Electronic date interchange is, of course, the most recently implemented and the smallest part by far. It is an interesting phenomenon that for intracompany messaging, the sharing of information between and among various departments and divisions within an organization, companies have often developed fairly sophisticated methods of saving information and sending messages. For example, many companies have databases that can be accessed randomly using keys that point the requester directly to the information needed. Also, personnel are often linked together via E-mail systems, voice mail systems, ad hoc reporting systems, and executive information systems. There may even be links to outside sources of information such as news services, Dun & Bradstreet profiles, and industry updates. These provide the structure within which those in need of information can access the most current version as quickly as possible. When we examine the format of an EDI transmission, we see a simple, sequentially organized structure, with no capability of randomly accessing any particular transaction or piece of information. The sophistication of databases is nowhere apparent in the EDI data stream. It is up to each receiving organization to save incoming EDI data in such a way that they can easily be offered to all those in need of them.

Components of Electronic Commerce

The concept of electronic commerce has grown as a response to the need of business partners for an uninterrupted flow of information both within and between their companies. Electronic commerce is simply conducting business using a combination of electronic media to facilitate the flow of information. For example, companies are beginning to automate the generation of EDI transactions out of structured E-mail and fax messages. They are also using CAD/CAM design blueprints to retool their manufacturing processes. Below are discussions of some of the electronic messages of which electronic commerce is comprised.

Narrative-Style Messages

Narrative or free-form messages communicated electronically via E-mail and fax have existed as an intercompany messaging capability for many years. However, contrary to EDI which is machine(computer)-readable, the free-form format requires human interface at each end. While transmitting these messages does save mail time, they must still be read and interpreted by people and, therefore, cannot be handled in an automated way.

CAD/CAM Drawings

In manufacturing, companies have been using computers to aid in the development of designs for component parts for many years. More recently, the completed designs have been transmitted intercompany in their file form rather than in a printed copy. This has enabled both the company defining the specifications and ordering the product, as well as the company designing and manufacturing it, to see and sign off on the exact same design. What's more, some manufacturing companies have installed software that allow them to prepare for the manufacture of a new product by retooling their machinery configuration directly from CAD/CAM designs.

Catalogs and Other Support Messages

While the number of business documents traded between companies is huge, the additional information needed to set up trading relationships and then conduct day-to-day business is equally large. Continuing to use paper business forms, lists, catalogs, and telephone calls to augment EDI transmissions places a tremendous burden on both the sending and receiving companies. As part of electronic commerce, companies are seeking to eliminate even these paper documents in favor of electronic transmissions of data. An interesting case is the catalog, which is typically very heavy in information content. Take, for example, the UPC catalog. This catalog contains the universal product codes of all the products offered by manufacturers in the grocery and retail industries. While all companies need to access a UPC catalog occasionally to obtain the product codes of particular products or product lines, the cost of manually accessing and maintaining catalog currency is very high. Unfortunately, the cost of down-loading the full catalog each time current information is needed is prohibitive as well. So some service providers are offering the UPC catalog on-line. By sending an EDI request for specific information, the requesting company can receive in a machine-readable form just the limited information that they need. Other types of support information such as trading partner profile information, contracts, and new store setups, are excellent candidates for electronic transmission as part of a more extensive electronic commerce implementation. However, today these types of information are either shared via fax, mail, or telephone conversations.

Is Now the Time for EDI?

It is obvious that there is tremendous interest in EDI today, as evidenced by the large numbers of companies participating in standards organizations, trade shows, and EDI classes and seminars. In addition, the number of publications specializing in EDI or containing EDI articles as they relate to a specific industry or type of company continues to grow every year. Today, almost all computer, purchasing, business, and industry magazines regularly feature articles on the benefits of electronic trade.

Whether or not EDI is right for you depends on several factors, such as

- Your industry's stance on EDI
- Your trading partners' pressure on you to implement it
- Your own business and systems environments

Understanding the business and technical issues surrounding EDI, analyzing the potential costs and savings of implementing EDI, and planning for and implementing EDI are what this book is about.

Summary

This chapter has provided the background knowledge of what EDI is, how it has developed, and how it can benefit your corporation. Electronic data interchange makes possible the transaction of business by using, in an automated way, machine-readable data transmitted from one company's business application to that of another. It permits companies to save both time and money because the data are transmitted rapidly, require no key entry, and have been generated accurately by a computer program. Consequently, it greatly reduces the order-cycle–pay period, which in turn allows a company to reduce its inventory levels and improve cash flow.

By examining the types of business environments that provide fertile ground for electronic trade, we can understand why certain companies and industries implemented EDI years ago, and we can anticipate the interest of others in the same or similar situations. By examining the business characteristics of documents that make them good candidates for conversion to EDI, we can see why so many companies have focused on the same small group of EDI transactions in their implementations.

While the business and system environments vary widely from company to company, within any one industry the business issues that exist are fairly homogeneous. Therefore, most companies within an industry choose to support the same transactions, those that promise to alleviate their primary business issues and concerns. For example, all the major automotive manufacturers send material releases against a precontracted blanket order to order parts and they receive advance shipment notifications (ASNs) to alert them of the impending shipment. In the retail and grocery industries, customer organizations are sending purchase orders, and receiving invoices and ASNs. In the pharmaceutical industry, wholesalers are sending purchase orders to place orders with manufacturers and charge backs to request their credit on sales to large hospitals who have negotiated special prices directly with the manufacturer. Some manufacturers are returning invoices and bid award notifications to alert wholesalers of the special hospital contract price.

Before implementing EDI, you will want to evaluate your various EDI opportunities and prioritize the transactions you intend to convert. Most probably you will begin with the same group as others in your industry. Again, the long-term benefits come, not from what you do, but from what *else* you do.

3
Are You Ready for EDI?

Assuming that EDI can be beneficial in many business situations, the question is, are you ready to implement EDI and take advantage of the savings it has to offer?

Objectives

- Describe why and how EDI is introduced into a company.
- Identify the minimum requirements of an EDI system.
- Identify the pros and cons of distributed and centralized processing.
- Identify and describe the four-step cost justification approach for investments in EDI.

How EDI Is Introduced

Typically, EDI is introduced at the operational management level of a functional business area. Sometimes, the idea stems from internal inefficiencies. This has been the case for many major customer organizations, specifically retailers and large manufacturers. Most often, the push toward EDI is caused by external pressure. This has been the case for the first-tier vendors to the major customers mentioned above. Typically, for every 100 EDI-active companies, 5 to 10 are customers and the remaining 90 to 95 are vendors. However, just by virtue of the fact that EDI is introduced at the middle management level, as opposed to farther up in the corporate structure, its first hurdle is encountered early. Typically, operational level managers are charged with accomplishment of departmental objectives using a predetermined number of resources for a specified time period, such as a year. When a new requirement or opportunity is encountered, this level of manager rarely has the authority to change direction or to expend more resources. So, in order to implement EDI, this manager must elevate the issue and sell EDI to his or her superiors. Because EDI is usually thought of as a technical initiative, and therefore sold internally as a technical initiative, upper management is not impressed or often not even interested. Probably, the most oft repeated message in this book will be

> EDI is not a technical project! While there *is* a technical component to it, it is predominantly a business initiative and requires companies to change the way they currently conduct day-to-day business.

Now that that's been said, we will get the technical requirements of implementing and supporting EDI out of the way before discussing the business aspects.

Minimum Requirements of an EDI System

As discussed in Chap. 1, only minimal technical capabilities must be in place to physically accomplish EDI. There are three main technical components for an EDI implementation:

- Intercompany communication capability
- A computer
- EDI software, application link, and translator

Following is an examination of each component with minimum requirements of each and the options from which you may choose.

Intercompany Communication Capability

In order to effect communications, both the sending and receiving parties must have a *modem* that enables their computer to communicate information over telephone lines. In addition, both the sender and receiver of a communicating pair must have the same communication protocol software that governs the communication of data between two points.

The protocol is designed to recognize which party is the controlling one (usually the sender), examine data that have been received, and acknowledge their receipt prior to calling for more data (usually in a unit called a *block*). If the protocol is error-detecting, the receiving side will be able to tell whether or not it has received data that are identical to those sent. When it detects that it has received incorrect data, it will call for a resend of the same data in the form of a negative acknowledgment (a NAK). If it ascertains that the data received are identical to the data sent, it will call for a send of the next block in the form of an acknowledgment (ACK). Note that the protocol makes no value judgments on the data themselves; it just verifies that data received are identical to those sent. Most companies consider error detection as the minimum communication capability for EDI.

There are error-detecting protocols for micro-, midrange, and mainframe computers. Traditionally, mainframe computers utilized bisynchronous protocols, while microcomputers used asynchronous ones. Today, there are both kinds for all sizes of computers. However, bisynchronous protocols can typically handle transmission of data at faster line speeds (baud rates) from 2400 to 9600 bits per second (where one character is equal to 8 bits), while asynchronous protocols handle from 300 to 4800 bits per second.

When dealing intercompany, compatibility of communication capability is a requirement. Both protocol and line speed must match that of your trading partner. As you can imagine, this can become quite an issue as your number of EDI trading partners grows. Not even all asynchronous or bisynchronous protocols are compatible with one another, so the protocol to be used for transmission of data between trading partners must be evaluated and agreed to for each trading partnership.

Often companies do not currently support and do not wish to support the variety of communication protocols needed to trade electronically with several trading partners. They may opt to transmit their EDI data through a company known as a third-party provider of EDI services or value-added network (VAN). This type of company acts as an intermediary between senders and receivers. It supports a large variety of communication protocols and line speeds and allows its customers to access the network using the communications configuration of their choice. Figure 3-1 illustrates how each partner uses the best communication scenario (protocol and line speed) between itself and the third party. This eliminates the compatibility issue between a company and its trading partners.

Also related to communications is the *operational window*, the number of hours per day and days per week that transmissions can be sent between trading partners. As opposed to mail service, which arrives ready or not, a receiving company must be ready to receive EDI data or the transmission will be unsuccessful. Value-added networks help in this area as well. Instead of limiting EDI transmissions to the hours in

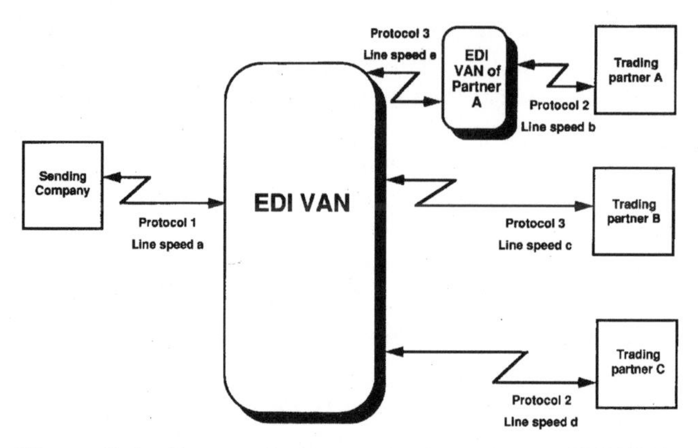

Figure 3-1. Communication scenarios using an EDI VAN.

which we are both ready, we can use the VAN as an electronic mailbox, which accepts data 24 hours a day, 7 days per week. As an EDI receiver we can then call in at a time of our convenience and retrieve data that has been sent to us.

While some companies transmit data directly between themselves and their trading partners, about 70 percent of EDI is sent via VANs. Value-added networks will be discussed again in a later chapter.

Software Requirements for EDI

As mentioned earlier, every EDI sender and receiver must have an EDI translator. There are various options you have when selecting the computer on which this program will reside. In this section we will examine the benefits and shortcomings of placing your EDI translator on either a micro-, midrange, or mainframe computer.

The microcomputer configuration is the least expensive but usually the slowest of the three choices. This is a possible solution for one of three types of companies.

1. The company that has no computer today but needs to do EDI and would use the microcomputer as a stand-alone computer
2. The company that already utilizes a microcomputer for its business application and wishes to utilize it for EDI as well
3. The company that has a midrange or mainframe computer but wishes to use a microcomputer as a front end for EDI

1. The Stand-Alone Microcomputer Solution. This configuration allows the company to physically accomplish EDI but offers very little in the way of benefits except elimination of mail float. The only way to generate an EDI transaction in this type of configuration is through key entry into preformatted computer screens. The only way to access information is by printing a paper document. This solution offers no link to an existing computer application. It not only does not eliminate manual tasks, it actually complicates them because users must get used to manually handling business documents in a whole new computer format. It also does not eliminate errors but may give users the opportunity to make even more. Obviously, this is a solution only if the company does not already have a computer application to which it can link its EDI transactions. Even then, it is only a short-term solution as it makes very little sense to implement EDI and not plan on implementing a system to generate or to process the EDI data.

On the other hand, when a small company is pressured by its major trading partners to implement EDI in a very short time frame, the microcomputer stand-alone solution may be the only one that they can actually accomplish with limited resources, time restrictions, and no technically oriented staff.

2. The Microcomputer Solution Integrated with Microcomputer Applications. If a company already has a business application on its microcomputer, then it may elect to implement a front-end EDI translator on its microcomputer. Visually, we can describe the business application as the one that acts as a door or gateway through which EDI leaves or enters the company. For example, a purchasing application is a gateway application for outgoing purchase orders as it is the one that generates the purchase order document; an accounts payable application is a gateway for incoming invoices as it is the one that reads and evaluates them. In this case, the translator would also reside on the microcomputer. On the sending side it would generate EDI transactions from a file of business information, not by key entry. On the receiving side, it would generate a file of business information from an EDI data file, not print out a business document. This sometimes is referred to as an *integrated solution* because the EDI data is integrated directly into the application environment with no intervening manual steps.

While the front-end microcomputer solution is also user-friendly and fairly inexpensive, it offers many more benefits than the stand-alone solution, mainly because it eliminates key entry and makes it possible to support higher volumes of EDI activity with a minimum of manual effort.

3. The Microcomputer Solution as a Front End to Midrange or Mainframe Computer Applications. The microcomputer can also be used as a front end to a midrange or mainframe computer. Even when companies have their business application programs on these large computers, they sometimes choose the front-end microcomputer solution for the short term to support the few EDI trading partners and little EDI activity they anticipate. This case is very similar to the previous one, except that the application resides on a different computer than the translation software. So, in this case, on the sending side, the application file would be down-loaded from the midrange or mainframe computer to be fed into the microcomputer translator. On the receiving side, the translator's output file would be up-loaded to the larger computer before being fed into the application program. While this is an inexpensive short-term solution, it often becomes cumbersome to manage when a company's EDI activity increases, either through the acquisition of more EDI trading partners and/or the introduction of additional EDI applications.

4. Midrange or Mainframe Computer Solution. The most efficient long-term solution for a company with a midrange or mainframe computer is to have both application programs and EDI translator residing on the same computer. With this configuration, the company can get the most feature-rich translator and eliminate any intermediary up- or down-loads.

Typically, midrange and mainframe translators support all the variable-length EDI standards and offer a great deal of flexibility in processing. In other words, the user can program the translator to vary its processing and decision making based on several parameters such as who the trading partner is and what transaction is being processed. It is just this capability, flexibility of processing, that allows you to respond to varying EDI requirements by your many trading partners without the need for custom programming on your systems.

Midrange or mainframe computer translators are considerably more expensive than those that reside on a microcomputer. In addition, they usually require technical expertise to install and maintain. However, if you are a midrange or mainframe computer–oriented company, the additional processing capability you get, and the benefits of having your translator contiguous with your application programs, make it cost-effective to make the investment.

Application Link Software

As mentioned earlier, there are two pieces of software generally required to link an existing business application to machine-readable EDI data; one is the translator, the second is the application link.

For the sending application link, development is minimal. Your main task is to identify the data requirements of the EDI standard and then determine which pieces of information you must collect and pass to the

- Business Application functions
  - Access internal applications
  - Accept key-entered data
  - Collect information needed to generate the outgoing business document
  - Process information
  - Generate outgoing business document
- Application link functions
  - Collect information from business application
  - Format information into a fixed-length computer file
  - Pass this computer file to the EDI translator
- EDI translator functions
  - Read fixed-length file
  - Use this file information to generate EDI standard transaction sets
  - Generate valid EDI standard file
  - Pass EDI standard file to communication facility
  - Maintain control information

Figure 3-2. Functions of a *sending* business application, application link, and translator.

translator in order to fulfill those requirements. This is primarily a technical task. Figure 3-2 shows the functions performed by the sending business application, the application link, and the translator for outgoing data.

Developing the application link on the receiving side will be a much more complex and time-consuming job. As mentioned in Chap. 2, there are two main functions of the receiving application link. The first is to feed a full transaction at a time to the receiving business application. For this function, you must first identify the data requirements of the business application and then collect the appropriate pieces of information to fulfill those requirements. Again, this is a technical task. The second is to replace, with automated processing, all the manual tasks that were previously performed on incoming business documents before their information was entered into the receiving business application. While this is a discussion about EDI software, you may be surprised to learn that specifications for the application link must be defined by the business people who currently perform the manual tasks. In fact, unless the business people take an active part in defining the required functionality of the application link and the logic needed to perform its procedures, the technical people have no way of knowing what the application link should do or how it should be done. We will be discussing in the next chapter the method one would employ to identify the requirements of the application link. This is no trivial task. Most companies that have chosen to automate their manual activities report that this is the most time-consuming and expensive part of their EDI implementations. See Fig. 3-3 for an illustration of the roles of the incoming EDI translator, application link, and business application.

The next question that is always asked is: "Should we purchase or develop our EDI translator and application link?" The short answer is purchase the translator, develop the application link. Here are some of the reasons why and the tradeoffs between development and purchase.

The translator should be purchased. One of the main reasons for buying it is to minimize or even eliminate the need to use your own internal resources for development. A second is that it distances your company from the EDI standard, which is complex and ever-changing. The

- EDI translator functions
  - Read and interpret incoming EDI standard file
  - Validate for correctness and completeness based on adherence to EDI standard syntax rules
  - Use this file information to generate fixed-length file
  - Pass fixed-length file to application link
  - Maintain control information
- Application link functions
  - Read fixed-length file
  - Perform editing and validations on business transactions
  - Pass a full transaction at a time to the business application
- Business application functions
  - Read and process full business transaction
  - Pass transaction information to internal files and subsequent business applications
  - Generate paper documents where necessary

Figure 3-3. Functions of a *receiving* translator, application link, and business application.

Develop	Purchase
Product specifically handles all of your company's needs	Product may have to be modified to handle company's requirements
Development must be scheduled according to priority of the project	Product already exists and is programmed to accept and generate EDI standard
Staff must have expertise in standard syntax and data requirements	Company need only produce the fixed-length file, not the EDI standard
Company must maintain product as standard changes	Regular updates to the product are produced and distributed by the software vendor
Development of the product must be scheduled and paid for using internal resources	Product is developed and requires no development staff resources
Implementation must be delayed until product is complete	Implementation can be accomplished as soon as the company has developed application link

Figure 3-4. Tradeoffs between developing and purchasing an EDI translator.

tradeoffs between developing and purchasing the translator are shown in Fig. 3-4. Today, virtually no companies develop their own EDI translator as there are excellent, cost-effective translators on the market for almost all brands of computer.

Now for the application link. Because its main purpose is to interact directly with your existing business application, which may have been developed and/or modified in-house, your best bet is to develop this program in-house as well. However, today many business application vendors are developing application links for their products. Sometimes they also develop alliances with specific EDI translator vendors. If your application vendor has developed such an alliance, you may want to take advantage of the convenience and improved ease of implementation by purchasing the EDI translator with which they have such an alliance. This will typically give you access to the application fields from which you must draw data or into which you must store data. It may also provide the exact fixed-file layout that the translator is expecting or use the exact file that the translator produces, with no additional work on your part. However, the part of the application link that must provide a substitute for your manual tasks must still be written in-house.

Distributed versus Centralized Processing

Let's begin with a discussion of the trends in corporate organization. Early on, business was conducted by manual procedures and was handled on a localized or distributed basis. In a small organization, there may have been staff members that wore many hats, fulfilling multiple functions such as receiving and processing purchase orders, checking inventory levels, directing shipment of goods, acknowledging receipt of the order and intention to ship, purchasing replenishment stock or component parts for manufacturing, sending invoices, receiving invoices and reconciling them to the purchase orders sent, and making or receiving payments. Larger companies had segregated staff so that different transactions were handled by different personnel. To the extent that the company had only one location, information was accessible to anyone who needed it, although it was often hidden by an intricate filing scheme.

As computers became popular, information (soon to be referred to as data) was "filed" or stored in centralized files on a corporate mainframe computer. However, in so doing, access was removed from the user. Jobs were typically scheduled and run by a centralized or corporate operations staff. Printed output from the various systems was disseminated to the user community.

Over time, the corporate operations department proved to be a bottleneck. Response to the user's request to develop and run new systems was slow, and printed output often took a long and circuitous route back to the requestor. To alleviate the bottleneck, corporate files were often divided up and distributed to individual users to give them access to the files that pertained to their functional area. Often, functional users were also given the ability to initiate and run jobs on the mainframe computer through remote job entry. Printouts of reports were routed to the local user. The drawback was that user job entry points were usually just terminals, with no computing capability of their own. And, the only systems that local users could run were those that already existed on the mainframe computer.

With the growth of the microcomputer and with the tremendous amount of memory and computing capability available at the user site, users wanted more than just access to their files. Even non–computer-savvy users wanted to store and process their own files on an as-needed basis, and they wanted to develop new applications locally. Consequently, local files replaced corporate databases of information. Systems became much more user-friendly, usually menu-driven and able to be run and managed by even a computer neophyte. The trend was for local departments to buy or develop software for their own use, with little regard for intracorporate processing or data access needs. This was an excellent solution for the business users who needed timely information and ad hoc reporting capability. However, it did not always serve the corporation well. It often led to the loss of ability to view data on the corporate level, to share information across department and division lines, and to support corporate files with a common support organization. Over time, many corporations developed what we sometimes refer to as *islands of automation*, highly automated systems that do not talk to other business applications and whose files cannot be passed to and used by other departments throughout the organization.

Enter EDI! In spite of the many good reasons why an organization chose to disburse processing and control of information to the functional area level, it turns out that disbursing the implementation and support of EDI is *not* a cost-effective and efficient choice. In fact centralized implementation, with one corporate gateway for all EDI activity, allows a company to save in every respect: dollars, resources, and time. A company that centralizes its EDI function and builds a corporate EDI gateway, need only

- Support *one communication link* between itself and all its trading partners or EDI VANs
- Purchase and support *one EDI translator* for all EDI activity throughout the organization.
- Fund one support organization to handle all EDI issues between itself and its trading partner base

The Best of All Possible Worlds

In order to provide access and control of information to the local user, while maintaining corporate databases and central processing capability, many companies have developed a mixed configuration of

- Microcomputers on individual users' desks onto which local files can be down-loaded
- Local area networks to provide a link between the microcomputers of coworkers and various departments
- Corporate midrange or mainframe computers which retain full databases of corporate information and possess the ability to process them

Often, these companies have attempted to standardize on software products used at the various remote locations throughout their organization. Sometimes they reduce their development costs by designing and developing a business application in a centralized corporate development group and distributing copies to the various functional locations throughout the organization. In the long run, this further reduces their EDI implementation costs because it allows them to develop only *one application link* to bridge the gap between their EDI translator and the common internal business application used by their various internal users.

While EDI has been used primarily to replace paper business documents, such as purchase orders and invoices, there is tremendous potential for using both EDI and other electronic commerce means of sharing business information to replace the multitude of telephone calls, voice mail messages, paper reports, and lists that augment those paper documents. If we think of electronic commerce as a solution to the information-based problems we are facing in business today, then we can broaden our scope and implement new and innovative applications that will ultimately provide big payoffs in our organizations.

Cost-Justifying Investments in EDI

The question is always asked:"How do you cost-justify investments in EDI?" Because EDI implementations are typically front-loaded with costs, while potential savings are longer term in nature, without cost justification many companies are choosing either not to implement EDI at all or to implement just enough to physically accomplish it without positioning themselves to attain any benefits. Either position is a mistake in the short term and potentially calamitous in the longer term.

Following is a four-step approach for cost-justifying an investment in EDI. See Fig. 3-5 for the four steps. Let me begin by saying that while this is a simple process, it is not necessarily easy to accomplish. It requires the time and active participation of the functional business community and an in-depth analysis of the current business and technical environments. However, the work expended is more than justified when you take advantage of the results of the study and implement those applications that have high payback potential. Let's discuss how to accomplish each step.

1. Analyze your current costs.
2. Determine which tasks you will automate.
3. Analyze and compute the EDI and automation-related costs.
4. Compare the two and develop a phased cost saving analysis.

Figure 3-5. Four-step approach to cost-justifying investments in EDI.

Step 1. Analyze Your Current Costs

There are three components to this step:

1. Identify the five-stage cycle in the life of the document.
2. Identify all manual and system tasks performed during each of the five stages.
3. Determine the cost of each task and the accumulated cost of all tasks.

1. Let's take a look at the various stages of a paper document. This is important because the average company grossly underestimates its paper processing costs as a result of underestimating the tasks involved. As a matter of fact, most often the analysis of tasks goes no further than listing the tasks of only one stage, that of initial processing for an incoming document, or generation of the document for an outgoing one. So, as a precursor to identifying tasks, let's describe the stages in the life of a paper document.

   For an incoming document the five stages are: receiving and distributing, initial processing, filing and redistributing and accessing of information, reconciling to other documents, and historical filing and ongoing access to information. For outgoing documents, the five stages are: initiating and generating the paper document, shipping and filing of the document, accessing document information, reconciling to other documents, historical filing and ongoing access to information.

   Appendix A contains a sample list of tasks for each of the five stages for both an incoming and outgoing document. This is not meant to be an exhaustive list of all possible tasks but rather a sampling of the types of tasks you can expect to find in each stage.

   You will notice that each stage is comprised of both manual and systems tasks. In most cases, unless careful study is done, only a small percentage of these daily tasks are identified. Often those asso-

1. Receive and distribute.
2. Perform initial processing.
3. File, further distribute, and access information.
4. Reconcile to other business documents and files.
5. File for historical storage and ongoing access to information.

Figure 3-6. Five stages of an incoming paper document.

1. Initiate and generate paper document.
2. Ship and file the document.
3. Access document information.
4. Reconcile to other business documents.
5. File for historical storage and ongoing access to information.

Figure 3-7. Five stages of an outgoing paper document.

ciated with normal processing are listed, but those required to handle rarely occurring situations or error conditions, *exceptions,* are ignored or just not thought of. This is particularly dangerous because the processing of exceptions requires between 5 and 10 times the time and resources of normal processing. To make matters worse, it has been found that 1 in every 20 pieces of paper has a material error in it, so exception processing is often a sizable portion of our everyday work.

2. Since the actual cost of handling and processing a paper document is a compilation of the time and resources we expend in handling it during each of its stages, it is particularly important that we identify all the tasks. Most companies tend to underestimate these tasks. Therefore, the following steps are given to help you to identify them. This is not a management level job; it is best done by the employees who actually perform the tasks.
   a. Ask employees to identify and list the tasks that they perform during a predetermined time period such as a week or month. Encourage them to identify and write down even those that they perform only once or twice in a long while.
   b. Using this list of tasks, develop an information flow diagram to illustrate the flow of information through the functional business area as the tasks are performed. This diagram will also show instances where information is passed to people and systems outside of this functional area as well as the outside sources of information used in performing the tasks. A description of how to develop an information flow diagram along with a sample is shown in App. B.
   c. Once you have developed the information flow for your functional business area, ask the staff to check it for accuracy.
   d. While you probably won't clutter up the flow diagram with exception processing, be sure that it is amply described in your

list of manual tasks. Particularly note the logic that your staff uses to identify an exception; who, outside of your organization, they work with to make the corrections; and what outside information sources such as telephone, fax, price catalog, list of product codes, and reports they use to make the corrections or completions.

Likewise, you will need data flow diagrams of the systems used in this functional area. This is a system equivalent of the information flow diagram developed for the manual tasks. Your technical staff will most likely already have a system data flow that shows all the tasks the system currently performs. If not, the procedure to develop it is the same as shown in App. B. This task should be done by IS people very familiar with the system.

3. Finally, determine the costs of performing the tasks you have identified. You can compute a cumulative dollar figure by using your task lists for each of the five stages. In order to do this, you will need to answer three questions for each task:
   a. Who is performing the task? Develop your answer in terms of dollars for salary and benefits.
   b How long does it take? Do this for each task, based on the cost of the employee performing it, and then accumulate over all the tasks.
   c. What are they doing? Develop a detailed description of each task, how many times it is done per document, and for how many documents it is done in a typical period (day, week, month, year).

   Once you have determined the cost for each stage in the life cycle of the paper document, add them all together to compute the overall cost for one document. Then, multiply the overall cost by the number of paper documents you send or receive in the specified time period to arrive at the cost to your company of that document for the period.

   You will want to factor in the cost of exception processing and internally generated errors. For exception processing, try to estimate the percent of times that it occurs. For example, are 10 percent of the incoming orders incomplete? Do 20 percent need to have your product code substituted for the customer's code? Once you have estimated the correct percentage for a task, multiply that number by the number of paper documents in the period to develop a cost for the exception task.

   For internally generated errors, you will be interested in another statistic. On average, 3 to 5 percent of keystrokes have been found to be in error. Some companies report even substantially higher error rates. And these errors may have far-ranging implications. For example, incorrect entry of a product code from an incoming order, if not caught and corrected, can result in an incorrect shipment, which in turn may require pickup and restocking of product, and an incorrect invoice as well. The final result is most certainly an unsatisfied customer and late payment.

You will probably be amazed at how high your cost to manually process a business document is. Just as verification of your calculations, an oft-reported cost for an incoming purchase order is between $50 and $75. Be suspicious if you arrive at a figure substantially below this level for your incoming paper document, unless you have already automated a great deal of the previously manual tasks.

Step 2. Determine Which Tasks You Will Automate

To accomplish this task, you will go through an exercise of rethinking your functional area procedures in general. Using the business and system flow diagrams that you have developed, determine which of the tasks would be good candidates for automation. The key here is to select first those tasks that are people-intensive, paper-intensive, rote, easy to develop system logic for, and using information that is already in machine-readable form. See Fig. 3-8 for a list and description of each.

Just a note of caution here; don't wait to automate until you believe you can computerize 100 percent of the cases in a process. This is usually close to impossible. Your typical business situation will be comprised of a certain percentage of cases, say 80 percent, that are fairly easy to automate, while the remaining 20 percent require more sophis-

- People intensive

  Using large quantities of your most expensive resource
- Paper intensive

  Using many copies of the same and associated paper documents. Often the most error-intensive tasks as well
- Rote

  Repetitive. Usually not requiring the expertise of people
- Easy to develop system logic for

  Accomplished with little difficulty. Decision based on the answer to three to five questions

Figure 3-8. Characteristics of highly automatable tasks.

ticated logic, outside information, or expertise difficult to transfer to a computer application. To segregate these cases, consider developing a selection module that precedes the automated process and directs the easy ones to a computer program for processing and the difficult ones to an experienced person for handling. By automating 80 percent of your traffic, you will experience substantial savings and at the same time free up your people to spend more time on the difficult 20 percent. Your staff will then be able to accomplish new tasks that they may never have had time to tackle before.

By the way, the 80 percent easy, 20 percent difficult example usually follows fairly closely to the real situation. The message here is to understand what tasks are being performed. If you go after the least sophisticated, easiest to automate tasks first and phase in more automation over time, you will begin to realize savings in the short term with a minimum of development and implementation costs.

Step 3. Analyze and Compute the EDI-Related Costs You Expect after Automation

Following are discussions of the six categories of automation-related costs.

1. *Programming costs.* For each set of tasks that you select to automate, you may need to design and cost out the development of programming logic that will be housed in up to four computer programs described below.

1. Programming costs
   Developing systems that perform previously manual tasks
2. Purchase, development, and support of an EDI translator
3. Selection of and contracting with an EDI VAN
4. Computer and support costs
5. Education
   Providing the training needed for executives, business management and users, and MIS managers and staff
6. Working with potential EDI trading partners

Figure 3-9. Six categories of automation-related costs.

a. *Selection and routing module.* Each time you choose to automate only a percentage of the possible cases of a particular task, plan on developing a selection and routing module. This program will separate those to be handled by the computer application from those to be still handled by a person and will pass the former to the application program described in Part *b* below and the latter to the appropriate person. This module is actually needed even when you plan on automating 100 percent of the cases because you will phase in EDI partners over an extended period of time.
b. *Processing program.* This will be a new program or enhancement of an existing program that will perform the automated tasks. Using the detailed description of the task, include in this program the logic needed to perform the tasks. Give it access to the information needed to accomplish its tasks as well.
c. *On-line exception processing module.* This will be a user-friendly, on-line module that will allow those people who need to be kept in the information loop to handle processing exceptions related to a transaction and then to return them to the natural processing flow as quickly and efficiently as possible.
d. *On-line information access module.* An on-line module for those who no longer have access to the paper documents. This new module should provide access to business information for those that need it to perform their jobs effectively. If you design this module to fulfill your business users' needs for information, it will become their link to machine-readable EDI information. With ease of use and random access to needed information, it will be more valuable and easier to access than paper documents ever were. This on-line tool alone will do more to sell EDI to your business user than all the "canned" EDI benefits that we all know and love.

2. *Purchase and support of an EDI translator.* Every EDI-active company must have an EDI translator to provide the link between the EDI standard and your internal files. Depending on your business and system environment, you may opt to install your translator on any computer platform, micro-, midrange, or mainframe computer. Of course, cost varies in direct proportion to the size of computer you select. Microcomputer translators can sell for as little as $1000, while mainframe computer translators can sell for as much as $50,000. In addition, there is typically an annual license renewal fee of anywhere from 10 to 15 percent of the price of the translator that entitles you to regular product updates and tables containing new versions of the standard. Even though in-house development is a possibility, translators are particularly complex programs and require a tremendous amount of expertise in the standard. Few, if any, companies

choose this route today. Instead, they select from the many excellent, cost-effective products on the market.

3. *Selection and contracting with an EDI VAN.* Seventy percent of EDI traffic flows through an EDI VAN. Most companies contract with a VAN as soon as they begin implementing EDI. While there are charges associated with VAN services, off-loading your communications support to an outside company greatly limits your company's internal operations support needs. In addition, VANs typically provide robust security and controls as part of their service. These same levels of security and controls would be expensive to install and support at an individual company site.

   There are a few noticeable exceptions to the overall trend of using EDI VANs. Walmart is one such exception. Walmart already supported its own communications network for the transmission of intracompany information and already had an operational staff to support it. They have opted to support EDI communications through this same facility. However, for most companies this would not be a cost-effective alternative.

4. *Computer and support costs.* Once you have implemented EDI and automated your manual tasks, you will find that you are using a greater amount of computer resources and require specialized operational personnel to support EDI, communications, and enhanced computer applications. In a start-up or limited EDI implementation there may be only one person working part time on EDI. In a multi-application, aggressive EDI implementation, there may be several people dedicated full time to EDI. Typically support requirements grow over time in direct relationship to the amount of business conducted electronically. This is not necessarily an additional cost to your company. Often it is more of a repositioning of resources. There is typically a sharp decline in the number of key-entry and paper-handling clerks needed and an increase in the number of more technologically skilled personnel. People requirements during the planning, implementation, and support stages of an EDI program will be discussed in more detail in Chap. 6.

5. *Education.* The most often mentioned reason for lack of success in EDI is *lack of education.* You may immediately see the need to educate technical people on various computer-oriented EDI topics; however, the most often mentioned barrier to the growth of EDI is *lack of commitment and buy-in* by top executives and functional business people. For this reason, we recommend education of all levels of personnel, specifically geared to their areas of concern and their need to know. Below is a sample list of educational topics.

   a. Executive overview of electronic commerce

b Developing a corporate EDI and electronic commerce strategy
c. What is EDI and how does it work?
d. The business environment needed to support the electronic environment
e. The technical environment needed to support EDI
f. Operation and support of EDI software and communications
g. Auditing and controlling the electronic environment

There are many educational options from which to select; on-site consultative education programs, public classes, industry and EDI trade shows with educational workshops and vendor displays, educational videos, and EDI and industry publications. There is even an on-line information service for EDI professionals with a bulletin board feature that you can use to network with other EDI users and learn from their experience.

6. *Working with potential EDI trading partners.* Not too long ago, one of the largest costs of implementing EDI was the cost of traveling and meeting with prospective EDI partners to explain EDI concepts and convince them to do it. That is no longer the case. Virtually all companies have heard about EDI. The vast majority believe that the question is no longer "Will we do EDI?" but "*When* will we do EDI?" Many are already in the process of planning for EDI or are already EDI-active with other trading partners. However, trading partner–related EDI costs have not disappeared. There are still many issues to discuss; some business-oriented, regarding how business will be conducted in the electronic environment, and others technical, regarding communications and standard issues. Some companies even conduct trading partner conferences during which they describe their current and future directions for EDI and outline the technical specifications of their programs to many partners at one time.

Some of the costs mentioned in this section are one-time costs that you can expect to incur at the beginning of your implementation. Others are ongoing costs that can be included in your post-EDI monthly cost estimate.

Step 4. Compare the Costs and Savings for a Phased Implementation Approach

The final step then is to compare pre- and post-EDI costs and estimate the savings you can expect in your post-EDI business and system environment. Typically, a company new to EDI goes through a period during which EDI is a cost item to them and represents no savings. This can be for a negligible amount of time or for an extended period depending on two parameters; how much up-front analysis, design, and application development precedes their implementation, and how quickly they add partners and convert business to EDI. The most common reason why EDI remains a cost item over an extended period of time is that it is implemented as a stand-alone technical capability, not as a tool to be used in conjunction with newly streamlined and automated business procedures.

However, even in the most aggressive EDI programs, there is a start-up period that is heavily front-loaded with costs for education, purchase of software, and development of business applications. After this initial stage, the company is ready to phase in EDI with one trading partner after another. Depending on the amount of business actually converted to EDI, savings will vary.

Here is a sample cost saving analysis based on hypothetical figures. The estimation of savings is purposely conservative and the estimation of costs is purposely liberal so as not to skew the results unrealistically. Following are the assumptions made:

- The company receives 1000 purchase orders per month.
- The pre-EDI cost of handling and processing each order is $60.
- The company plans on automating virtually all manual procedures related to incoming purchase orders. The expected cost of new software development is $240,000.
- The post-EDI cost of handling and processing each order is estimated at $6.
- The company chooses a mainframe translator whose price is $50,000 plus an annual renewal fee of $7500 after the first year.
- It takes 6 months to complete software development and prepare for the first EDI partner.
- The company converts 10 percent of its paper purchase orders to EDI by the end of the initial 6-month period and increases its EDI activity to 20 percent 3 months later, to 30 percent 3 months later, to 40 percent 3 months later, to 50 percent 3 months later, and then stays at 50 percent.

Figure 3-10 illustrates the document handling and processing costs this company can expect as it converts from manual to automated processing. At the beginning, with no business handled by EDI, the total monthly cost of $60,000 is attributable solely to paper processing. Whereas, when 50 percent of business has been converted to EDI, the combined EDI and paper costs are only $32,500. This represents a monthly savings of $27,500. The computations have purposely not been carried any further than 50 percent EDI because this is a reasonably

% EDI	$: EDI portion/mo	$: Paper portion/mo	Total $/mo	Savings/mo
0	0	1000 × 60 = 60,000	60,000	0
10	100 × 6 = 600	900 × 60 = 54,000	54,600	5,400
20	200 × 6 = 1,200	800 × 60 = 48,000	49,200	10,800
30	300 × 6 = 1,800	700 × 60 = 42,000	43,800	16,200
40	400 × 6 = 2,400	600 × 60 = 36,000	38,400	21,600
50	500 × 6 = 3,000		33,000	27,000

Figure 3-10. Document handling costs—paper and EDI.

Category	One time/Repeating	$
Develop application software, $40,000/mo.	6 month project	240,000
EDI Mainframe Translator	One-time	50,000
Education—General and Software	Classes on time	12,000
Communications	Per month charges:	250/month
	10% EDI	300/month
	20% EDI	400/month
	30% EDI	500/month
	40% EDI	600/month
	50% EDI	
Translator renewal contract	Once/yr	$7500
Time spent with trading partners	Repeating	2,000/yr

Figure 3-11. Additional costs.

attainable goal.

Figure 3-11 shows additional costs; some one time only, others ongoing.

The graphs in Figs. 3-12 through 3-14 illustrate the cost analysis of this company over a 2-year time period. Each shows progressively more detail of the same analysis.

While this type of analysis requires time and effort to produce, it is particularly powerful as a selling tool for EDI and for the reengineering of your internal procedures. In fact, what you will find is that the cost of implementing only EDI is low but provides little in the way of savings. On the other hand, the cost of reengineering and automating of tasks, while substantial, provides commensurately larger savings.

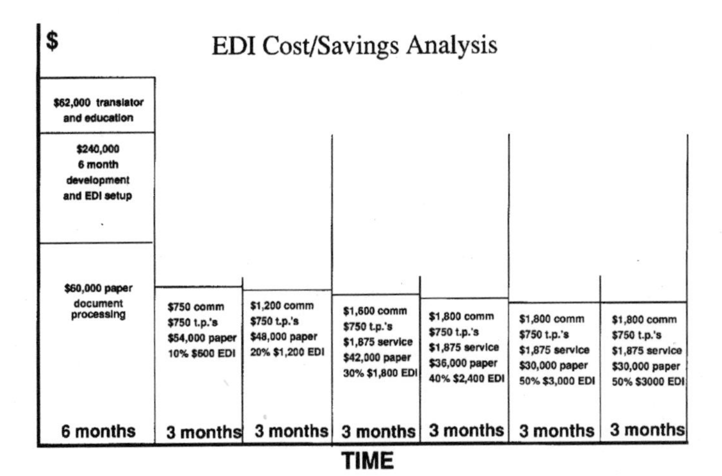

Figure 3-12. Cost analysis—3-month figures over a 2-year period.

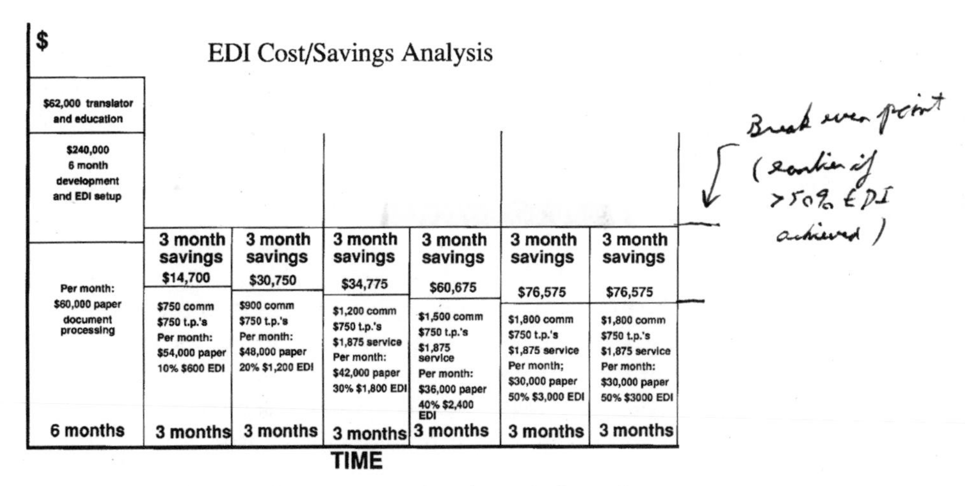

Figure 3-13. Cost analysis—Accumulated quarterly savings.

Summary

Technical readiness for EDI is easy to attain as the amount of new technology needed for EDI is minimal and inexpensive; the data files transmitted between trading partners are simple in format, the communications requirements are basic, and translation software is available for almost all brands and sizes of computers. As a matter of fact, EDI can be

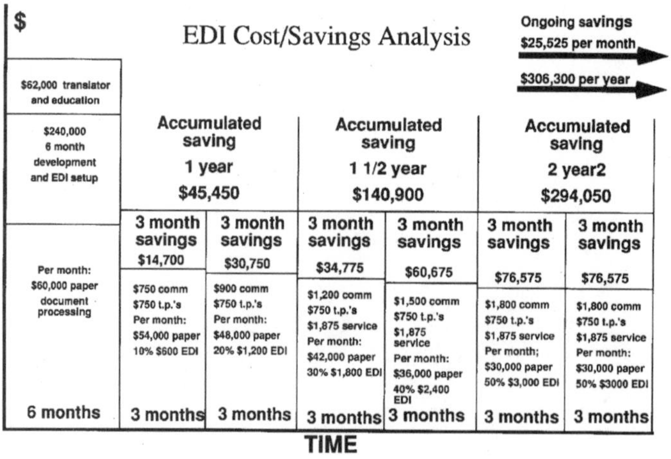

$
EDI Cost/Savings Analysis
Ongoing savings
$25,525 per month
$306,300 per year
$62,000 translator and education
$240,000 6 month development and EDI setup
Per month: $60,000 paper document processing
6 months
Accumulated saving 1 year $45,450
Accumulated saving 1 1/2 year $140,900
Accumulated saving 2 year2 $294,050
3 month savings $14,700
$750 comm $750 t.p.'s Per month: $54,000 paper 10% $600 EDI
3 months
3 month savings $30,750
$900 comm $750 t.p.'s Per month: $48,000 paper 20% $1,200 EDI
3 months
3 month savings $34,775
$1,200 comm $750 t.p.'s $1,875 service Per month: $42,000 paper 30% $1,800 EDI
3 months
3 month savings $60,675
$1,500 comm $750 t.p.'s $1,875 service Per month: $36,000 paper 40% $2,400 EDI
3 months
3 month savings $76,575
$1,800 comm $750 t.p.'s $1,875 service Per month; $30,000 paper 50% $3,000 EDI
3 months
3 month savings $76,575
$1,800 comm $750 t.p.'s $1,875 service Per month: $30,000 paper 50% $3000 EDI
3 months
TIME

Figure 3-14. Cost analysis—Accumulated annual savings and ongoing savings.

introduced in even the smallest PC-oriented companies. In reality though, benefits are attained only when more than EDI is implemented. So, readiness for EDI has little to do with EDI at all. It is more a matter of readiness to change your business culture, readiness to rethink your intracompany business procedures, readiness to provide a systems environment that supports automated decision making and transaction processing, in short readiness to change.

The typical entry point is at the middle management level or lower within a functional area. From this point, word about EDI, both benefits and requirements, needs to be disseminated upward to top level management, downward to business users, and sideward to the Information Systems (IS) organization that supports the business applications for this functional area. Again, openness to change is almost a requirement for success in this venture. When implemented properly, EDI brings about changes in all parts of an organization and for all people within the organization.

Even though it is almost intuitive that cost savings will accrue from the introduction of EDI and automated processes, investments are rarely made on intuition alone. In order to gain the support and funding for an aggressive EDI program, most organizations require a cost-justification exercise that quantifies both the anticipated costs and savings. For this study, determining what is happening today is all-important. While this undoubtedly will be a short-term drain on resources, in the long term it is what provides the basis for computing post-EDI savings.

4

Biggest Payoff Applications of EDI

Objectives

- Identify main document types in seven categories of business transactions.
- Identify the main EDI applications for a retailer, manufacturer, wholesaler/distributor, and financial institution.
- Trace the intracompany flow of information for a manufacturer.

Categories of Business Transactions

For purposes of this discussion, business documents have been divided along functional lines into seven different categories (see Fig. 4-1), and a sampling of document types has been included in each. These are not meant to be exhaustive lists of all documents used to transact business. They certainly do not even represent all the documents for which an EDI standard format has been designed. You will find that in your industry there are specialized documents that fulfill business needs that may not even exist in other industries, and these may prove to be your best EDI opportunities. However, these lists do contain many of the documents that are being used the most in EDI today and others that may or may not become EDI favorites in the future. Keep in mind that conversion of a paper document to an EDI format merely provides the transmission vehicle. In order to automate the use of the electronically

1. Administrative
2. Prepurchasing
3. Purchasing
4. Shipping and Receiving
5. Warehouse
6. Customs
7. Billing and Paying

Figure 4-1. Categories of business documents.

transmitted data, you must first upgrade the computer application to perform the manual tasks associated with the document and you must reengineer business procedures to remove people from the normal processing loop. Exceptions will still be handled by people.

1. Administrative

The administrative category is somewhat of a catchall for documents that fall outside of day-to-day business needs. Included are some of the documents that are used as backup documentation for business transactions, and others that can be used to help us make wide-ranging business decisions. Examples of administrative documents that are good candidates for use in EDI format are:

- *Product code and price catalogs.* In EDI format, these provide current product information that can be included in outgoing purchase orders, verified against incoming invoices, and used in store and shelf configuration systems.
- *Catalog updates.* Maintaining the currency of catalog information is extremely important. Updates received in electronic form can automatically replace existing, out-of-date items or be added as new items.
- *Forecasts and plans.* For vendors, receiving forecast information from a customer can be particularly useful when planning their own manufacturing schedules and when purchasing from their own vendors. In the automotive industry, this transaction alerts a vendor to an increase or decrease in product demand. Sometimes, it is even used as an order itself.
- *Deals and promotions.* Special deals offered by manufacturers to their customers are often complex and require a great deal of manual effort

to evaluate. In the grocery industry, the large chains are requesting these deals in electronic format so they may automate their evaluation process. They expect substantial savings from making buying decisions on results of the automated analysis.

- *Statements.* In instances where purchase orders and invoices are both being handled via EDI, statements provide an after-the-fact reckoning of business activity for a specified period. Generated by the vendor, this transaction is used to automate the reconciliation between vendor and customer files. It can also be used as a summary from which customers can analyze their buying patterns.

2. Prepurchasing

As the name implies, these transactions precede the purchase of product. The examples included fall into convenient pairs comprised of an inquiry and a response document. They are not only useful in their own right but form the basis for the purchasing transactions as well.

- *Request for quote (inquiry document)—quote (response document).* These documents may or may not prove to be useful as EDI transactions. The request for quote (RFQ) document often is comprised of two components, one containing the basic business information and the other providing manufacturing specifications, product drawings, and quality requirements. The information component of the RFQ is fairly simple to convert into electronic format and use as input to a computer application. However, the auxiliary information, while able to be transmitted electronically, is difficult to handle in an automated way. They are included here because the federal government is beginning to require quotes in electronic form from their vendors.
- *Inventory inquiry—inventory advice (response).* These are excellent EDI transactions. The inquiry is easy to generate out of a manufacturing requirements planning (MRP) system, and the response is easy to read, understand, and integrate into a purchasing system. In fact, a favorable inventory advice can be automatically converted into a purchase order. A vendor can offer rapid response to inventory inquiries as a premium service for its customers.

3. Purchasing

Not only do the family of purchase order documents appear in this category, but others that also provide the information needed to effect a purchase appear in this category as well. The purchase order transaction is the most used EDI message.

- *Purchase order.* This transaction has many uses. The most obvious is as input to an order-entry system. However, interestingly, the purchase order contains 70 percent of all the information ever shared between business partners. With the purchase order already in machine-readable form, the receiver can easily use its information for several other purposes. For example, purchase order information can be used as input to a business system that examines and reports on trends in ordering patterns.
- *Purchase order acknowledgment.* This transaction acknowledges receipt of a purchase order and confirms for the customer how much of the order the vendor intends to fill and when to expect delivery. In some industries, such as grocery, it is not used because the shipment of goods arrives soon enough to preclude its need.
- *Purchase order change and acknowledgment of the change.* While order changes can certainly be transmitted via EDI, they are most often handled off line, meaning that they are communicated directly to a person either by phone or fax and handled as exceptions. Even when the change is handled manually, acknowledgment to the change can be generated and sent via EDI.
- *Material release.* A material release acts as the day-to-day ordering mechanism for product covered by a blanket or long-term purchase order. This is the ordering transaction typically used in the automotive industry. Because of the JIT manufacturing schedules maintained by automakers, vendors are expected to immediately respond to it by delivering product and generating and returning via EDI an advance shipment notification.
- *Point of sale—Inventory on hand.* These two transactions perform similar functions. One or the other is generated by the customer to provide status on its current stock position. Once received by the vendor, the transaction is automatically compared to the desired model stock plan of the customer and an order for the needed product is generated and shipped. We will be discussing this service known as quick response in the retail industry and as efficient consumer response in the grocery industry in Chap. 8.

4. Shipping and Receiving

Transportation-related transactions have historically been the mainstay of EDI. Until just a few years ago, tracking of rail cars and their contents was the largest EDI application by far. It was believed at that time to represent one-half of all EDI activity.

- *Shipment status inquiry—shipment status response.* Virtually every major trucking carrier today offers a shipment tracking system for its customers. Based on an EDI-formatted inquiry from a shipper, the carrier's system automatically responds with the most up-to-date shipment status information. The challenge is then for the customer to use the status information to automatically update pending receipts on their inventory files.
- *Advance shipment notification.* Today, this is the fastest growing EDI transaction. It is used to give advance notice of shipment product quantities and expected arrival time to the customer. It is one of the required EDI transactions in the automotive industry and is rapidly becoming a required transaction in the retail industry as well. This transaction has proved difficult to implement. One of the reasons is the EDI standard format itself, which was designed in a hierarchical form to relate information on all levels of the shipment. For example, on the largest level, the transaction describes the shipment, which may be a full truckload. Within the truck, the ASN contains identifying numbers and related information for each pallet. Within each pallet, it contains the identifying numbers and related information for each packing carton. Likewise, within each packing carton it contains identifying numbers and related information for each product. This is invaluable information for the customer but difficult to generate in an automated way by the vendor.
- *Bill of lading.* This transaction provides the complete product information needed to transport goods. Because it contains the same product information as the ASN, it can be generated as a by-product of the same automated procedure. This may provide extra incentive to develop a business application to generate shipment transactions.
- *Freight bill.* Another in the list of transportation-related transactions, the freight bill contains virtually the same information as the others. When a carrier receives an EDI freight bill describing the order it is expected to pick up and deliver, it can easily use it to compute the order volume and to generate its shipper invoice.

5. Warehouse

In the current business environment, accurate and timely information is rapidly becoming the norm. This means that an outside warehouse employed for storage of goods and shipment of orders must become an integral part of its customer's information loop. Warehouse transactions in EDI format allow the outside warehouse to get and send machine-readable data, just as if they were part of the shipping company. Some of the many warehouse transactions being used are

- *Inventory inquiry and inventory status.* Transactions related to inventory held by the warehouse
- *Shipping notice.* From manufacturer to warehouse to alert of an impending shipment of goods
- *Receipt confirmation.* Response from warehouse to manufacturer to confirm receipt of goods
- *Shipment order.* Copy of customer order to be sent from the warehouse
- *Shipment confirmation.* Confirmation of order being sent to customer

6. Customs

Prior to EDI, imported goods could remain on the dock awaiting release to the importer for an extended time period of up to 6 weeks. Using EDI transactions, information regarding goods being imported into the country arrives well before the goods themselves. The EDI customs declaration can be processed while the shipment is in progress. Often the customs release is prepared and ready by the time the goods arrive, allowing for instant release of imported product. The two customs transactions that have been converted to EDI formats are *customs declaration* and *customs release.*

7. Billing and Paying

The final category of business documents contains financial transactions. Both the invoice and payment remittance transactions are particularly useful as EDI transactions. When receiving these transactions in electronic format, companies can develop business systems to automate the invoice and payment reconciliations. These procedures are typically the most people-, time-, and error-intensive procedures within the organization.

- *Invoice.* This transaction contains most of the same information found on the purchase order. In addition, it contains actual shipping quantities, prices, and the invoice date. The design of EDI standard transactions facilitates the generation of this type of turnaround response transaction. This makes it easy for a company receiving an

EDI purchase order to generate and send an EDI invoice. The receiver of the invoice can attain substantial savings by developing a system to automate the comparison of the EDI purchase order and the invoice. Many believe that the receiving of an EDI invoice is the highest-payback EDI application in terms of potential dollar savings.

- *Payment remittance.* The payment remittance contains much of the same information as the invoice, plus status per line item of what is and is not included in this payment and why. This transaction contains detailed information describing a payment. The payment remittance transaction can be used for two purposes: the first is as described above to provide information about a payment, and the second is to authorize a financial institution to electronically transfer funds from one account to another. When used in the former role, it is typically transmitted from customer to vendor either directly or through an EDI VAN. When used in the latter role, it is transmitted to a financial institution, which acts on the funds transfer and then forwards the remittance information to the vendor's bank via the automated clearinghouse network between banks. The vendor's bank then transmits the payment remittance to the vendor. Even though there is no requirement that the payment remittance transaction be used only with electronic funds transfer (EFT), many companies have opted not to use it because they have made the decision not to pay electronically. Unfortunately, they are forsaking its main benefits, that of allowing the vendor to understand what is being paid for and to properly and quickly book the payment to the customer's account.
- *Credit and debit memos.* These transactions can provide the added information needed to eliminate reconciliation exceptions and completely automate the process. For example, when the customer relates via payment remittance that it is not paying for a particular invoiced line item because of a previous credit for goods received damaged, the vendor would normally process this manually as an exception. With EDI credits and debits already on file, the vendor's system would automatically find the credit, process it as part of the payment, and eliminate it from the outstanding credit file.

As mentioned above, this was not meant as an exhaustive list of business transactions. For other transactions prevalent in your company or industry, you may want to determine if EDI standards already exist. If so, you may want to consider using them. If not, you may want to become part of the ANSI X12 standards group that designs EDI formats for new transaction sets.

Using EDI and Electronic Commerce in Various Types of Companies

Let's examine the intercompany flow of information for various types of companies: retailers, manufacturers, distributors, and financial institutions. For each type there are several business areas that generate and send or receive and process business transactions to and from trading partners. Which areas are they, what transactions do they handle, and how will the current business environment be affected by EDI?

EDI in a Retailer

For a retailer, often the most pressing business issue is the handling of inventory: purchasing it, storing it, getting it on to the selling shelves, and tracking it through the selling process. As illustrated in Fig. 4-2, the outside entities (represented as squares) with whom a retailer deals are its suppliers, carriers, and banks. The gateway applications (ovals) are purchasing, accounts payable, and receiving. Because of this, the biggest EDI opportunities are to

- Send the purchase order
- Receive the invoice
- Receive shipping documents
- Send payments

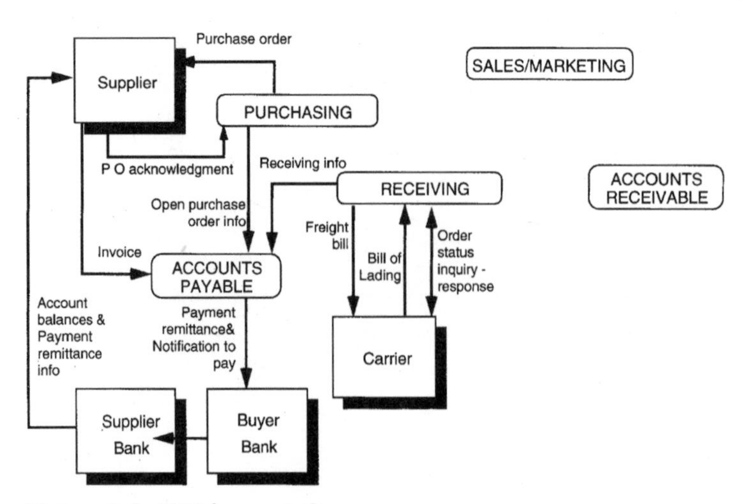

Figure 4-2. EDI for a retailer.

While most retailers have kicked off their EDI programs by sending the purchase order, it is in automating invoice reconciliation and in receiving goods and automating inventory file upgrades that their largest savings lie.

EDI in a Manufacturing Company

Figure 4-3 represents a manufacturing company. The outside entity trading partners, customers, suppliers, financial institutions, carriers, and public warehouses, are represented by squares. The functional business areas are represented by ovals. Transactions to and from trading partners are shown along arrows. In most organizations, these are paper-based or telephoned transactions, but potentially each could be an EDI transaction.

The same ordering and paying transactions occur with directions reversed between this company and its customers as are shown for the company and its suppliers. They would flow between the customer's purchasing area and the supplier's sales and marketing area and between the customer's accounts payable area and the supplier's accounts receivable area. They are

- *Preordering transactions.* Request for quote, quote, inventory inquiry, and inventory advice.

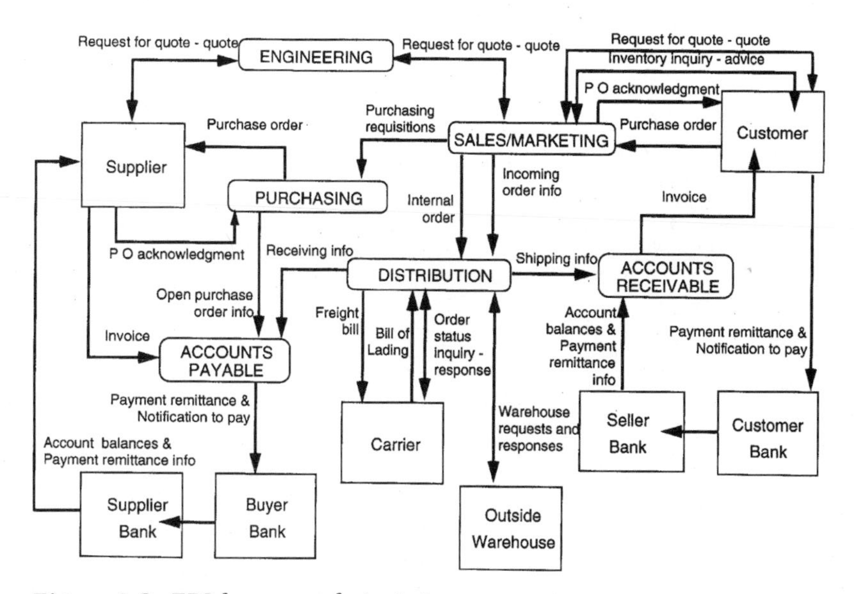

Figure 4-3. EDI for a manufacturing company.

- *Ordering transactions.* Purchase order, purchase order acknowledgment, purchase order change, purchase order change acknowledgment, order status inquiry, and order status response.
- *Billing transaction.* Invoice.
- *Payment transaction.* Payment remittance.
- Additionally, shipping transactions flow between the manufacturer's distribution area and its carriers, and shipment and inventory transactions flow between the distribution area and an outside warehouse.

EDI in a Distributor-Wholesaler Company

Figure 4-4 uses the same symbols to illustrate the outside entities, functional business areas, and transactions used by a distributor-wholesaler. Note that all the gateway business areas, those dealing with outside entities, are the same as for the manufacturer except for engineering and the outside warehouse. We show the same information flow as for the manufacturer as well, since the company deals with the same outside entities; customers, suppliers, carriers, and banks.

While EDI represents a potentially large opportunity for savings, actually implementing EDI with many trading partners is dependent on the size and sophistication of the trading partner base and their EDI-readiness. For example, the distributor whose customer base is com-

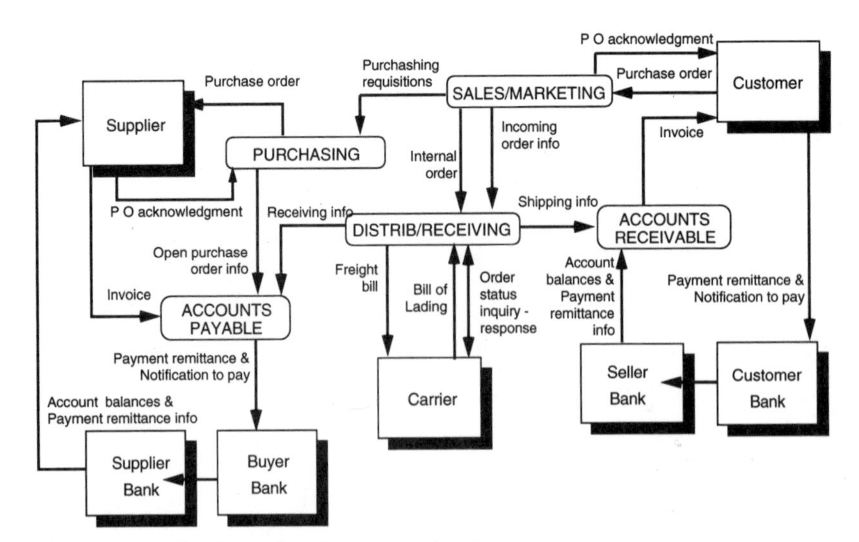

Figure 4-4. EDI for a distributor-wholesaler.

prised of small retail stores may not have a customer base interested in integrating EDI into their day-to-day business activities. However, one whose customers are large chain and mass merchandisers is most likely already being pushed to implement various EDI applications as a requirement for doing business. For these customers, the distributor may even offer additional services such as inventory inquiry and automatic ordering system. As we will discuss in Chap. 8, this vendor may offer a quick response program to its major customers, comprised of a combination of business services, EDI, and various other technologies. Because distributors-wholesalers are typically high-competition, low-profit-margin companies, they are quick to embrace new technologies that promise cost savings and provide the ability to offer higher levels of service. Electronic data interchange is one such technology.

EDI in a Financial Institution

Again, using the same symbology, Fig. 4-5 shows the transaction flow for a financial institution. Buyer and seller customers of the bank have been shown separately just to illustrate the flow of information. Also illustrated are bank suppliers and other financial institutions. Electronic data interchange potential exists in two main areas: one as a pass-through for customer payment remittance information and the other for sending purchasing-related transactions to a supplier and receiving acknowledgment and invoice transactions in return. Because a financial institution is purchasing in relatively small quantities compared to the

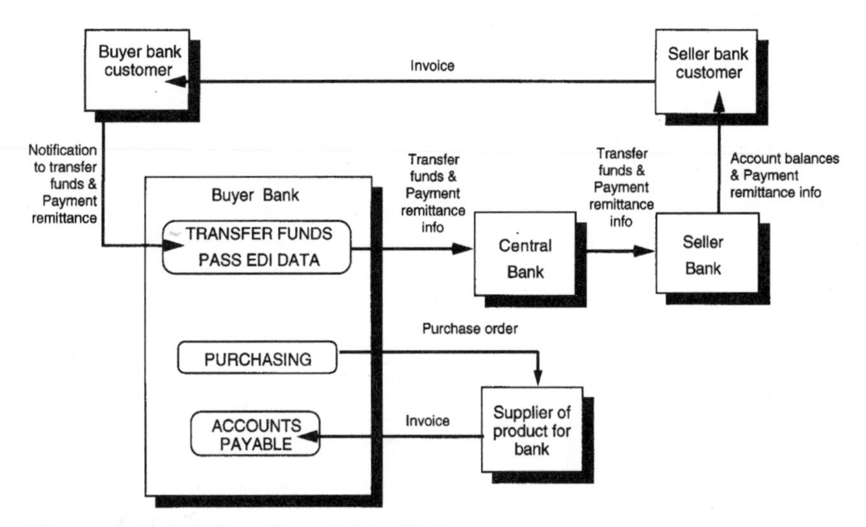

Figure 4-5. EDI for a financial institution.

manufacturer, who uses its purchases in the manufacturing process, and the retailer and distributor, who resell it, the financial institution is usually not a good candidate for receiving transportation transactions via EDI.

Intracompany Flow of Information for a Manufacturing Company

So far we've looked at the intercompany interfaces that exist between various types of companies and have discussed the EDI transactions that they might use. However, the key to realizing savings with EDI lies in the intracompany automated use of data received electronically. That is, passing of EDI data into and through internal applications with no interspersed manual interfaces. Using a manufacturing company as an example, let's track the flow of data from one business application to the next as they pass through the various functional areas.

Figure 4-6 illustrates the application environment of the same manufacturer as was shown in Fig. 4-2 with the external and internal data needs of each business application added. From this you can see how EDI data are used throughout the organization. When the various business systems are upgraded to accept computer files and perform the previously manual tasks, there is no need for additional data entry.

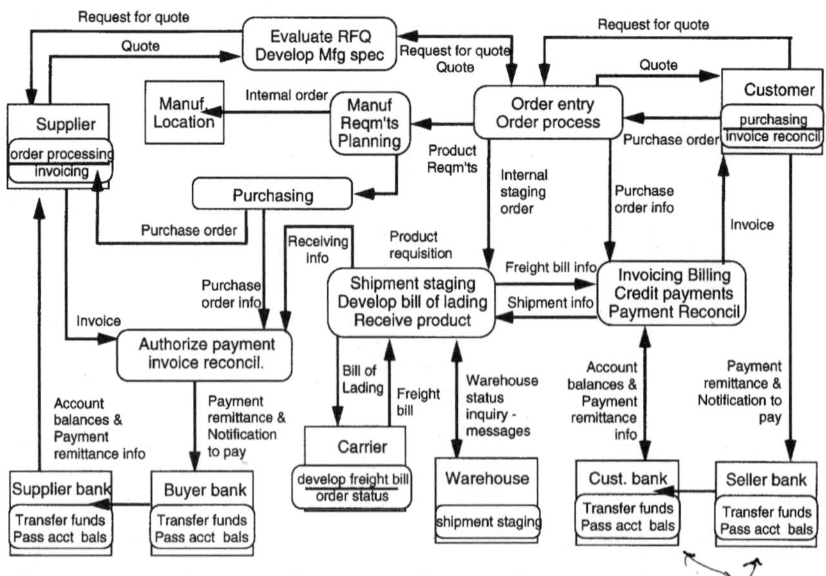

Figure 4-6. Application environment of a manufacturing company.

Let's follow the flow in the figure, starting with an RFQ received from a customer, either for stock or custom product. If for stock product, the quote responding to the RFQ can usually be generated directly out of the order-entry application just by examining on-hand inventory in the inventory file. However, when the RFQ is for custom product, the data may need to be passed to the engineering department, where they are evaluated, used to develop manufacturing specifications, and finally responded to. Implementing a completely automated system to accept, interpret, and act upon RFQs is difficult because

- Such a request is usually accompanied by drawings and other textual information, which may need to be handled by a person or evaluated by another system.
- The receiver of the RFQ may need to query potential suppliers of raw materials and component parts to develop a response to the RFQ.
- The responding quote may be a complex transaction containing price, manufacturing schedules, anticipated product availability, and delivery schedule.

Figure 4-6 shows integrated order processing and MRP systems. With such an integrated system environment, this company stands to benefit greatly from the use of EDI.

Let's examine the internal flow of information, beginning with the purchase order, which is received into the sales and marketing area and passed to the order-processing application. Customer information on the order is compared to customer information maintained on the customer file. When a mismatch is found, incoming data are used to update information already on the file. Customer credit is then checked and the transaction proceeds. Next, the system determines if the order is for stock or custom product. If the order is for custom product, it is directed to the appropriate person for manual processing. However, if it is for stock items, the order is directed through the order-processing system which determines if enough inventory is available to fill it. Regardless of inventory position, it passes the order along to the MRP system for its evaluation. If stock is available, the order-processing system generates an internal staging order which it directs to the warehouse for staging of the shipment. If not enough stock is available to fill the order, the MRP system generates an internal order to produce more product. The MRP system also examines the stock position on raw materials and component parts needed to manufacture the product. If low, based on predefined inventory requirements, the MRP system generates a purchase order for replenishment of the needed inventories. Finally, the order-processing application determines if an acknowledgment is needed. If so, it generates a purchase order acknowledgment using much of the same information included in the purchase order.

After determining if stock is available for shipment and handling shortfalls as agreed to by the trading partners, the order-processing system sends a duplicate of the internal order used for staging the order to the billing application. The billing system initially generates the invoice to match this order. However, if updated information is received from the shipping department which changes shipping quantities, the invoice is changed to reflect actual shipping quantities. Order processing also passes a copy of the order to the payment reconciliation application. When payment is received, this application reconciles it against purchase orders, invoices, and shipping information and posts it to the customer's account.

By upgrading the gateway application to accept machine-readable data and perform previously manual editing tasks, this manufacturer is able to pass incoming purchase order data to its various other systems: MRP, shipment picking and staging, billing, and payment reconciliation. Not only that, but the manufacturer can use this same information as input to its purchasing system for raw materials and component parts needed for manufacture. All of this is done with no additional key entry or human intervention.

Let's continue by following the internal order generated by the MRP system. It is passed to the warehouse shipment staging application where it automatically generates the pick instructions given to the warehouse person. Any discrepancy between quantities to be picked and actual quantities found in the warehouse are used to adjust the invoice. Subsequently, actual shipping quantities are sent to the billing application which uses them to generate an accurate invoice.

The warehouse uses shipment information when it deals with carriers. Another business system generates the freight bill for the order. The freight bill is then transmitted via EDI to the carrier who rates the shipment and returns a freight invoice. If freight charges are passed along to the customer, then they are the last component of the final invoice. Whether freight charges are paid by the customer or shipper, the freight invoice is passed from the warehouse to invoice reconciliation for carrier payment.

Finally, when payment for the order is received from the customer, the payment remittance transaction is reconciled against the invoice and the dollars are used to quickly and accurately update the customer's account. Even discrepancies between the invoice and payment can sometimes be automated, if computer-readable credit and debit memos are accessible to the payment reconciliation application.

So, aside from adjustments to the internal order caused by discrepancies between the inventory file and the actual quantities found in the warehouse, there is no need for additional key entry or hard-copy documents throughout the entire order-ship-invoice-payment cycle. Of course, users need to have access to status information and to handle exceptions and reinstate them into the normal processing flow.

When internal applications are integrated, as in this example, the same accurate and timely data flows through all systems. In this example, the data received via EDI are passed from one system to another with no intermediary manual procedures and no chance of introducing errors during interpolation or key entry. This is just the type of system configuration that offers the biggest benefits from EDI.

In addition to all the above, incoming purchase order information can be used for a multitude of other applications. For example, order information can be put into systems that analyze sales trends, develop new marketing directions, and configure inventory requirements, just to name a few. Outgoing orders can provide input to systems that analyze procurement and set inventory levels.

The foregoing discussion covered business between the manufacturer and its customers. When that company trades with suppliers, virtually all the same transactions are used, but their directions are reversed. The manufacturer sends the purchase order, sometimes generated from its MRP system and sometimes through requisition from various internal departments. In return, it receives an invoice into accounts payable which it reconciles against an open purchase order and verification of receipt of product. Manual invoice reconciliations, notoriously wasteful in both time and effort, are eliminated as all three components are already in machine-readable form. A copy of the purchase order and the receiving information is collected by scanning bar-coded shipping carton labels. When the incoming invoice matches an existing purchase order and when product has been received, the payment can automatically be authorized. Only exceptions need human intervention.

Potentially, even payments can be automated. Instead of generating and mailing a check, payment can be made by transferring funds through EFT from the customer's to the vendor's bank account. When each of the trading partners deals with a different bank, the transfer is accomplished through the banking system's automated clearing house (ACH). Electronic funds transfer transactions can be authorized by means of an EDI payment remittance transaction. This same transaction provides detailed payment information to vendors to allow them to properly credit customers' accounts.

The manufacturer used in this example had a full complement of business systems that not only performed their own day-to-day business procedures but were integrated to share business files with one another as well. As you begin to evaluate your company's EDI potential, think about the business systems or lack of them in your company today and the plans you have to reengineer your procedures in the near term. The more robust your system environment, the more you can expect to get out of an investment in EDI. The fewer systems you already have, the more you should think of reengineering as a precursor to EDI.

Summary

This chapter has discussed the most promising EDI transactions. It has also examined the organization of several different types of companies and has traced business information flow through a manufacturing company.

Often the catalyst for introducing EDI into a company is outside pressure of a trading partner. Sometimes it is internal inefficiencies. In either case, business managers need to evaluate the readiness of their organization to proceed in light of existing business systems and communication capability. In general, the more business systems in place to provide automated processing and the more integrated they are to provide an uninterrupted flow of data, the better positioned a company is to implement EDI and then to use the machine-readable EDI throughout its organizations.

5 Business Issues of EDI

This chapter will discuss the business issues that must be resolved and agreed to by trading partners when they implement EDI.

Objectives

- Understand the impact of EDI on a functional business area.
- Discuss the growth of EDI standards and demonstrate how they have been designed to carry all needed business information and provide exceptional controls. Demonstrate the importance of supporting the EDI standard.
- Identify the security issues of concern when implementing an EDI system.
- Discuss the determinants of the decision to communicate directly or through a third party.
- Identify the various business arrangements agreed to by trading partners when implementing EDI.

Impact on the Business Area

As was mentioned, initial implementation of EDI usually encompasses the conversion of one or more business documents in a functional business area from paper-based to electronic format. So, the initial impact on a company is limited to the affected business area: its information flow, change in organization and job tasks, and required expertise levels.

In the pre-EDI business environment, the transaction of intercompany business is accomplished by sending and receiving paper forms which are designed to hold all the information required by the receiver to perform the business transaction. However, in practice, information is often missing or incorrect on paper documents, necessitating a good deal of interpretation and handling on the receiving end prior to system processing and taking action. Sometimes, errors are even introduced into the data at the receiver site during interpretation and key-entry steps.

In the current paper-based business environment, sending of a business transaction is a three-step process. One, some business trigger perceives the need for it; two, it is generated through a combination of manual and system processing; and, three, it is sent to the trading partner for processing. When you are an EDI sender, the impact on your organization can be minimal. The business trigger often is unchanged and generation remains the job of business people interacting with a computer application. The change to your business is limited to conversion of the hard-copy document to electronic form and replacement of the handling and mailing activities by transmission of the EDI standard file.

It is only when companies use the implementation of EDI as an opportunity to rethink their current business procedures and streamline and automate their day-to-day tasks and decision making that they impact their functional areas. One of the first challenges in automating a previously manual task is finding an automated way to kick off or trigger a procedure. For example, today generation of a purchase order is preceded by the sending of a product requisition to a procurement person. This requisition then acts as the trigger for the purchasing function. Often the requisition is in the form of a telephone call or paper-based memo. So, automating the purchasing function requires not only converting the paper purchase order into an electronic format, but converting the requisition into a computer file as well and probably developing a system that can be used by the various internal departments to generate their product requisitions.

With this system in place, the automated procurement function trigger is in place. However, the logic associated with how much is purchased and from whom and when the product must be delivered is still not resident in a system. The next challenge is to incorporate as much of this logic as possible into the purchasing system. Three things are required here. The first is cooperation of the existing staff members in defining what tasks they are currently doing and how they make decisions. The second is converting the outside sources of information they currently use to make their decisions to computer files that can be accessed by the newly upgraded purchasing system. The third is identifying the exception conditions that arise during the course of business and how they are detected, corrected, and reinserted into the normal business flow.

As you can imagine, the impact on the staff of automating these procedures is great. At first glance it appears as if purchasing jobs will disappear to be replaced by an automated system. Can you really expect your purchasing staff members to happily define their job tasks so they can be replaced by a computer application?

In practice what actually happens is that job tasks change. The rote aspects of the purchasing jobs are eliminated as people tasks are converted to computer tasks. This provides spare time in which these same people can perform new tasks that are often even more valuable to the organization. For example, instead of entering purchase order information, the procurement person can evaluate new product lines, study vendor service levels, and determine who to purchase product from for the most timely and cost-effective service.

While EDI is often viewed as a threat, it often represents an opportunity for those affected. With elimination of rote tasks and an increase in new tasks, employees are positioned to make better decisions and become more productive. It is not only worthwhile but often mandatory that the company provide training for affected employees to prepare them to perform new job tasks and become familiar with new automated job tools. When change is handled properly, it can be a positive experience; when it is not, it can lower the morale of your staff and reduce productivity.

I have just used the purchasing function as an example. The same could be said about other sending applications. If we merely convert an outgoing invoice to electronic form, while keeping the manual interfaces status quo, there will be little impact on the functional business area. However, if we streamline the billing function and eliminate the people interface to make way for automated generation of the invoice, we will experience similar effects on the staff of the accounts receivable function.

Let's look at the effects of EDI implementation when you are the receiver. For companies that receive an EDI data stream, produce hard copy, and handle paper documents, there is little impact on jobs and people. However, there is little benefit as well. For those that take this opportunity to rethink and reengineer their internal procedures, the impact can be great. Take sales people whose primary function is currently to handle incoming telephone or mail orders. With the elimination of the paper or phone order, and the automation of manual editing, virtually all the order-related tasks are handled by a computer. Does that mean that all the sales people are let go? Certainly not, although the staff may initially believe that this is what's in store for them! However, it does mean that the sales staff will have time to *sell*. Now, this may seem like an obvious opportunity for the company. However it is, only if the sales people actually know how to sell, which may not currently be the case as the existing sales people were hired just to take orders. Again, training may be required to handle changing job tasks. This leads to a more productive business environment in which incoming orders are handled more quickly, efficiently, and correctly, and sales staff is out finding new customers and selling more to existing customers.

Again, sales is merely an example. The impact is even greater in an accounts payable department, where the current staff is busy performing manual invoice reconciliations. This task can be almost completely automated when an organization matches an incoming EDI invoice with a copy of the previously sent EDI purchase order. Only exceptions need to be handled by people, so fewer accounts payable clerks will be needed. Existing employees may be trained to perform other jobs, often in other departments, and will most probably require retraining as well. As EDI is phased in, automating more of the day-to-day activities, the target size of the department is decreased. Getting to this target takes many forms. Sometimes, as employees leave for other opportunities, they are just not replaced. Sometimes, existing employees are used to fill head count requirements in other departments. Sometimes, there is a planned staff reduction.

Since implementing EDI is almost always accompanied by organizational change, anticipating the change and properly handling it should be a priority. How the organization will change must be addressed by top executives of your organization. Then, handling the change is done by midlevel management in conjunction with human resources staff. You may be surprised to hear that people issues are the biggest barrier to streamlining of current procedures to support EDI. From top executives on down, there is a reluctance to change. In Chap. 7 we will discuss, in more detail, barriers to entry into the EDI arena.

Supporting the EDI Standard

You may be wondering why the EDI standard is being discussed as a business issue. Surely, this machine-readable format is the concern of the technical staff. There are two reasons why business people are particularly interested in the standard. One, the standard must contain a place for every piece of information needed to transact business. Two, the standard must provide the controls needed to ensure that we can track transmissions and can ascertain that all the data that was sent was, in fact, received by the ultimate trading partner. Let's address each of these aspects separately.

Business Information in the Standard

An EDI standard contains a predefined format, called a *transaction set*, for each EDI document. Some EDI standards contain formats for many different transaction sets. Others are limited to just a few. The format can be described as a skeleton containing a predetermined location for every piece of information deemed necessary to conduct business. Formats have historically been designed by business people, not technical people. The standards designers have been familiar with the information requirements and business needs of their industry, as they have come from the staffs of those companies and industries. When a standard has the narrow focus of only one company or industry in mind, its format contains just those information requirements needed for that company or within that industry. However, when the standard has a broad interindustry or even international focus, representatives from many industries and countries participate in their design. The X12 cross-industry standard and the EDIFACT international standard have many more information fields and much more functionality than their industry-specific and company proprietary counterparts. Historically, as new information requirements are identified, the standard is enhanced to include them. As a matter of fact, there is a mandate that each standard must satisfy all information needs of the group of companies to which it pertains, so deficiencies are always corrected and updated versions of the standard are put into use. If, as you begin to implement EDI, you find that the standard you intend to use does not fulfill your information requirements, you may either evaluate other standards for a better fit or submit a change request to the designing and maintaining body of that standard. While you are not guaranteed to get the change exactly as you have requested it, you will definitely get a way of handling your particular business need.

Years ago, the standards were always in a large degree of flux and represented a major barrier to implementing EDI. Companies were afraid to begin using them. While they wanted to be on the leading edge, they were not interested in being on the "bleeding" edge.

Today, this is no longer an issue. EDI standards are widely accepted. In fact, the transaction sets that are already designed and in use are very complete and stable. However, more than one EDI standard exists. Each has been designed to fulfill the information requirements of a particular group of companies. Some are narrow in scope and were designed to fulfill the requirements for one industry only. Companies in an industry having industry-specific standards have typically chosen to accept and use that standard when they initially implement EDI. The beauty of industry-specific standards is that they are based on the homogeneity of business requirements across the industry. As such, they are focused on the transaction sets and specific information needs of companies within that industry. Some examples of industry-specific standards are

ORDERNET/NWDA, for the pharmaceutical industry

UCS, for the grocery industry

WINS, for the warehouse industry

EAGLE, for the hardlines industry (The hardlines industry is composed of hardware and houseware and includes everything sold in a home center.)

TDCC, for the transportation industry

When a company from one of these industries implements EDI, it can immediately trade with all its industry trading partners who have chosen to support the same industry-specific standard. However, these companies often choose to adopt the broader-focused X12 standard in the long term. They do this for two primary reasons. One, industry-specific standards such as ORDERNET and EAGLE were designed as fixed-length standards and do not contain compression efficiencies gained by eliminating extraneous blanks and zeros as does the more modern X12 standard. Two, they find the industry-specific standard too restrictive, either because it contains too few document types to fulfill their business needs or because they wish to trade electronically with partners outside of their industry.

When companies do business with partners in other industries, they need a more broadly accepted and used standard. The emergence of the ANSI X12 standard with cross-industry application has been a tremendous boon to EDI. The X12 standard is designed to fulfill the information needs of all companies and industries throughout North America. However, because of the large variance in information needed to support such a large and diverse user base, it tends to be much more complex and much richer in functionality than any of the industry-specific standards . Each X12 standard transaction set has been designed by business representatives from the various companies and industries that expect to use it. With such broad-based participation in its design, there has been a ground swell of acceptance for this standard from many different industry groups. Again, because of the broad base of existing X12 users along with prospective users from new industries, there is a constant need for maintenance of existing transaction sets and for development of new ones. Any company that finds that the X12 standard lacks a location for needed information or is unable to support its business needs can petition the organization to upgrade its transaction set to include the new information or functionality. The mandate of the ANSI ASC X12 organization is to satisfy the needs of all industries using their transaction sets.

As mentioned, the complexity of the X12 standard has led several industry groups to limit the exposure of their member companies to the standard by defining in an industry-specific implementation guideline the subset of the standard that they intend to use. If you are in one of the industries listed below, you will want to examine your industry guidelines rather than the full standard when converting a paper document to its electronic representation. Most likely you will find all your information requirements already defined within the subset for your industry. Some industries that have developed implementation guidelines are shown below along with the name of their subset.

Automotive industry	Automotive Industry Action Group (AIAG)
Chemical industry	Chemical Industry Data Exchange (CIDX)
Petroleum industry	Petroleum Industry Data Exchange (PIDX)
Electronics industry	Electronics Industry Data Exchange (EIDX)
Office products industry	Intercompany Communications Office Product Standard (ICOPS)
Retail/apparel industry	Voluntary Interindustry Communication Standard (VICS)

Controls in the X12 Standard

Because of the widespread acceptance of this standard, we will take the time here to discuss it in a little more detail. From the business point of view, it is important that the standard provide the ability to track EDI activity. An EDI sender knows what has been sent but needs to know what his or her trading partner has received. This leads to two basic control needs. One, the receiver needs to ascertain if he or she has received a full transmission, that is, everything that the sender sent. Two, the receiver needs a way of relating to the sender that the transmission has been received and whether it appears to be complete and correct according to standard syntax rules. There are mechanisms in X12 and several other standards such as UCS, WINS, and TDCC to fulfill both of these control needs.

Let's look at the various components of these standards to see how their controls work. The standard is comprised of a series of *transaction sets,* electronic representations of paper-based business documents. Included in a transaction set are all of the *data fields,* individual items of information, needed by the receiver to transact the indicated business transaction. Transaction sets are comprised of *segments,* an EDI term meaning records. Segments in turn are comprised of one or more related data fields.

Because business transactions are fairly homogeneous over industries, one transaction set is designed to handle the data requirements for all industries. However, because business systems vary in companies and industries, the standard transaction set must also have a great deal of flexibility. That flexibility is ensured by defining fields generically and giving each user the ability to use them in the way they are needed. The way this is accomplished is to precede a generic field with a qualifier code that tells the receiver how to interpret the meaning of the generic information. The standard provides a list of codes from which to select. For example, if one were to look for a field in an X12 transaction set in which to put "ship-to name" or "bill-to name," one would find only a generic *name* field. However, by preceding this generic field with an appropriate qualifier code, the sender relates that the name contained in the generic name field should be interpreted as ship-to name or bill-to name. To do this, the sender selects from the list of valid codes the one that is defined as the ship-to or bill-to name. For example, when the EDI sender places a ship-to name in the generic name field, he or she will select SH for the qualifier code field. The receiver then interprets the name found in the generic field as a ship-to name. Likewise, if the sender precedes the name with BT in the qualifier code field, the receiver interprets the name as a bill-to name. When an industry develops guidelines for usage of the standard, they typically examine the valid code lists and preselect those that are needed by their member companies. This serves two purposes; one, it simplifies the code selecting process, and, two, it eliminates code differences between trading partners.

Each ANSI X12 transaction set is described in a series of diagrams that illustrate the segments available for use in its header, detail, and summary sections. Transaction sets were purposely divided into the same three sections as their paper counterparts.

Figure 5-1 relates the electronic form of an X12 transaction set to a paper document. Information in the header area pertains to the entire transaction and may be overridden by information in the detail area. The detail area corresponds to the line item area of a paper transaction. The summary area is similar to the total line of a paper transaction but additionally contains summary information which is used as a control mechanism to assure that a complete transaction was received and certain fields, such as quantity, were not tampered with.

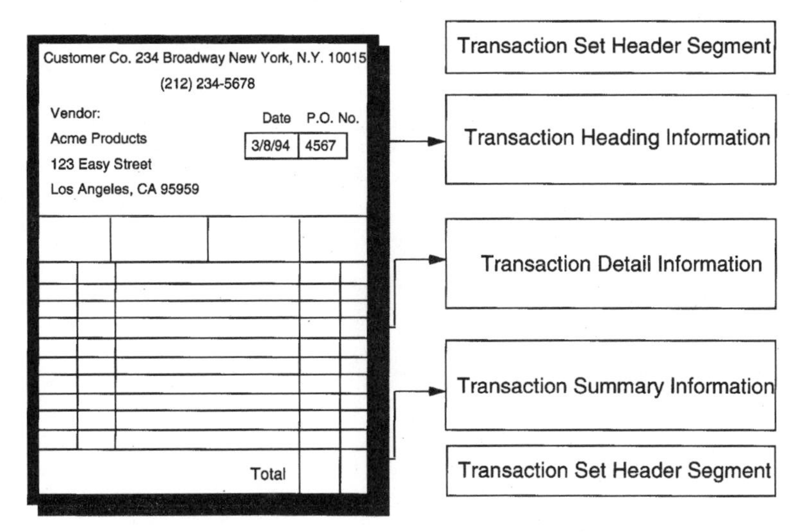

Figure 5-1. Relationship of a paper document to an EDI transaction set.

All the diagrams pertaining to a transaction set can be found in the X12 *Transaction Set Diagram* documentation. Each diagram contains a list of allowable segments for the transaction set and section represented. Segments are designated as mandatory or optional, can be repeated up to a predefined maximum number of times, and must be used in the order specified in the diagram. The same segments are used in many different transaction sets.

Just as transactions are composed of segments, segments are composed of data elements, whose usage designators may be mandatory, optional, or conditional. Conditional elements must be used in conjunction with one or more other elements as noted in the X12 *Segment Directory*. Elements are defined in the X12 *Data Element Dictionary*. Each has a name, identifying data element number, definition, minimum/maximum length, and data type (e.g., numeric or alphanumeric).

In order to generate an X12 standard file from a group of separate transaction sets, control segments must be added. Figure 5-2 illustrates the series of control envelopes required to build an ANSI X12 standard file. The outermost level of control is called the *interchange envelope*. It is defined as containing all data going from one sender to one receiver in a single transmission. It is delimited (directly preceded and followed) by an interchange header segment at the beginning and a trailer segment at the end. Directly inside of the interchange envelope is a functional group, defined as one or more transaction sets of the same type. It also is surrounded by a control envelope. In this case, a functional

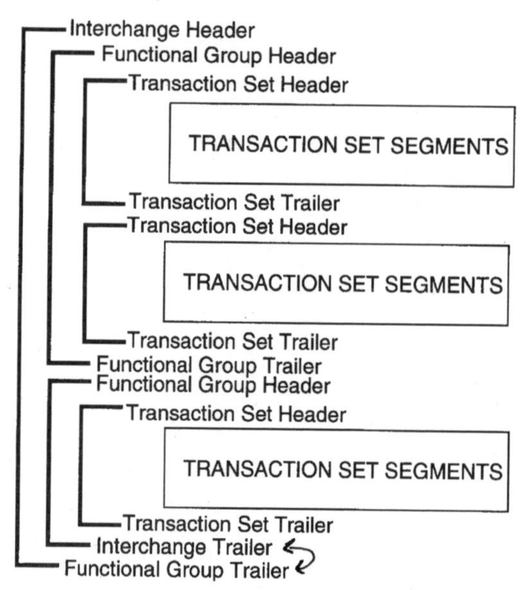

Figure 5-2. Standard control envelopes.

group header directly precedes the first transaction set of the group and a functional group trailer immediately follows the last transaction set. Finally, inside of the functional group are the transaction sets themselves. Each is delimited (started and ended) by a transaction set envelope comprised of a transaction set header and trailer.

Each header segment always contains an identifying number that uniquely identifies, for this transmission, the entity to which it applies. The header also contains other data fields needed for routing or interpreting the data to which it pertains. Its matching trailer always contains the same identifying number as well as a count of the number of units found in the envelope. An interchange trailer contains the count of the number of functional groups contained in the interchange. A group trailer contains the number of transaction sets in the group, and a transaction set trailer contains the number of segments contained in the transaction set.

Unlike the typical internal file, which remains in one place while being accessed by various internal users, EDI files are transmitted and require these control envelopes to provide the control and trackability needed by both the sender and receiver of the data. Specifically, when data are transmitted, the receiver must be able to verify that a complete transmission has been received. This is accomplished by the sending and receiving EDI translators. The sending translator generates all the control envelopes, identifying numbers, and control counts. The receiving translator reads the incoming EDI data stream, evaluates the control envelopes by looking for matching header and trailer identification numbers in each control envelope, and verifies that the number of units contained in each trailer is identical to the number it has counted. If the header and trailer identifying numbers do not match—for example, the receiving translator receives functional group header 1 followed by functional group trailer 2—it knows that the end of group 1 and the beginning of group 2 were lost during transmission. It rejects this incomplete group, and the receiving company calls for a resend of the data.

If the unit counts in the trailer and the number counted by the receiving translator do not match—for example, the trailer reports 10 purchase orders in the group but the translator only receives 9—the translator rejects the group and the receiver again calls for a resend.

To provide documentation to the sender of the results of the header-trailer reconciliation and examination for adherence to standard syntax rules, there is a standard transaction set called a *functional acknowledgment.* This relates that the receiver has received the transmission and has or has not found it to be correct and complete according to the syntax rules of the standard. When the functional acknowledgment is used regularly by EDI trading partners, its absence signals to the original sender that the receiver may not have received the data stream at all.

Security

Figure 5-3 lists several security issues of concern when implementing EDI. For one thing, there's no hard copy to file and distribute. Second, there's no place for authorization signatures in the EDI transaction. Third, how can a company be assured that the data sent were received and in the exact form as sent? Fourth, how can data be protected so they cannot be read by the casual hacker? Fifth, how can activity be tracked? Let's deal with each of these issues and see if there are ways of assuring the security of data while providing access as needed.

1. No hard copy to file and distribute
2. No place for authorization signatures
3. Need to verify that data were received exactly as sent
4. Data need to be protected from inspection by the casual hacker
5. Activity needs to be tracked

Figure 5-3. Five security issues in EDI.

Hard Copy

Since paper copy has been the traditional method of storing paper business documents, most companies have designed their business procedures around the use of the paper. Most business people are very leery, to say the least, of eliminating paper. Actually, paper is far from the ideal medium for storing and retrieving information as it is easily lost and subject to damage. In addition, paper takes up valuable space in office file cabinets and off-site storage facilities. Because of this, we try to save and file as few paper copies as possible. We typically try to determine how we'll want to use information in the future and then file documents so that we can access them in that way. For example, we might decide to file incoming purchase orders by customer name. We may even decide to keep a second copy and file it by date. However, the result of selecting only one or two ways of filing our documents is that we limit the ways we can access them as well. If, after the fact, we are interested in accessing and using document information in some other way, we have to rifle manually through all the documents to collect the information we need. For example, if our sales people want to conduct a sales campaign targeting customers who purchased particularly large quantities of a specific product in the last quarter of the year, they would have no easy way to do it. Or, if the marketing department wanted to compare purchasing patterns between last year and this one, they would be hard pressed to accumulate the information. And even if they did, the information would still be resident on pieces of paper and would still need to be key-entered before it could be used. I believe that, if it were not for the fact that business people are familiar with paper, it would be their storage medium of last resort. Finally, misfiling of a paper document can make it almost impossible to ever find.

In comparison, data storage is much more economical and secure as a storage medium. Even backup copies can be stored very economically and accessed within minutes. What's more, once data are in a machine-readable file, they can be stored simultaneously under multiple fields. This can be done by storing information in special files called keyed files or databases. The storage keys in these files are predefined, and business documents are automatically stored according to the information found in their key fields. This, in turn, makes the data available for random retrieval by any or a combination of the predefined storage keys. For example, I may ask to see all the orders I received between this date and that for specified quantities of one product or a combination of specific products. As long as a storage key is already specified at the time that data are stored, business people can use it to gain access to information. There is never a need to rekey information that is already stored. In fact, databases are not only keyed files, but they offer particularly efficient means of accessing information. One need only have the necessary program logic or query in place to access the information needed.

Being able to randomly gain access to information, from either a keyed file or database, provides the biggest benefit of EDI. While it would be possible to make multiple copies of paper documents and file them in a variety of ways, it would be almost impossible to preguess all the possible ways that you may want to use file information in the future and it would still be necessary to rekey information from paper copies in order to use it in another computer application. By giving our business people the flexibility of accessing incoming EDI information in all the ways that they need it, we are taking best advantage of the electronic medium and getting additional value from the same data. The question is never, "What can I use incoming EDI data for?" That is always obvious. For example, all companies make incoming purchase orders available to their order-entry applications and incoming invoices available to their accounts payable applications. The real question is "What *else* can I use these data for?"

Most organizations today are beginning to make information from computer files or databases available to personnel on a terminal or desktop computer instead of on paper reports. On-line programs provide an instant link to the most current information that business people need to conduct day-to-day business. From the executive examining trends in sales, to the customer service representative perusing purchase orders in response to customer queries, the information they need is available when and where they need it.

Of course, for business personnel to feel confident about conducting business from computer files, you will need to provide a stable and secure system environment. Incoming EDI information will need to be processed in a timely manner and stored and backed up automatically. Systems will need to be secured against tampering, and the computer will need to be up and available during all working hours. Both control and security will be discussed later in this chapter.

Authorization and Limited Access

As you can imagine, securing EDI data against tampering is extremely important. If there's one thing we can say about paper, it's that we can tell when we have the original and we can see if it's been changed. What's more, we can plainly see when a signature guarantees approval or authorization. Of course, we can provide the same degree of security in the electronic environment. However, we must use different techniques than we currently employ in the paper world.

First, it is important to keep the original. When EDI data are received, a backup copy should automatically be generated and stored in a file that can only be written to once. While this file can be accessed many times, the information on it can be confidently retained as an exact duplicate of what was received. This can be used as legal evidence just as an original paper document is used today.

Let's discuss the signature or authorization issue. While actual signatures can be scanned into computer files, this is rarely the solution used by companies doing EDI. Instead, when authorization is needed on an outgoing transaction, it can be accomplished by placing an authorizing code in the file from which the EDI standard is generated or in the standard file itself. Here's how it's done. Let's say that a buyer in a purchasing department can purchase up to a specified dollar amount under his or her own authorization but that purchases over that amount must be authorized by a supervisor. When the buyer generates an order, the system automatically computes total cost. When the total is less than the cutoff amount, the order is routed to the outgoing transaction queue to be translated into the EDI standard and transmitted to the trading partner. However, when the total is more than the cutoff amount, the transaction is routed instead to a supervisor's computer, and a message accompanies it to alert the supervisor that there is a transaction awaiting approval. With the supervisor's higher level of authority, he or she can provide the needed authorization. Once the authorization is received, this transaction is also translated and sent to the trading partner. Authorization information may or may not actually be transmitted, depending on the requirements of the trading partner.

Tracking of Data

When business data are transmitted outside of your organization, business people within the organization will want documentation of their progress between your shop and that of your trading partner's. Even though EDI transmissions are faster and more secure than paper documents traveling via mail service, business people remain more comfortable with the current system than with electronic transmission. Luckily, it is easy to track the progress of EDI data and then to report back to those interested. Following are some of the most asked questions and answers about the ability to track EDI data.

How Can We Be Sure That Data Have Not Been Lost or Changed during Transmission? The answer is that for a communications session to be successful, both sender and receiver must have identical communication protocol software that controls the communications session. When the protocol is error-detecting, the receiving system can tell whether or not it has received exactly what was transmitted by the sending one. Here is a very brief description in layperson's terms of how it works. Data are transmitted over telephone lines with the protocol software on one end taking control of the transmission and sending a block of data. The protocol on the other end positively acknowledges receipt to show it has received data identical to that sent and negatively acknowledges receipt and calls for a resend when the data vary in any way from the original. A predetermined number of negative acknowledgments (NAKs) will cause the line to drop. When the sender fails to get a successful end-of-transmission message from its protocol software, he or she knows that the data must be resent. When the receiver fails to get a successful end of transmission, he or she should purge the data already received and expect a resend of the full data file.

Because all companies doing EDI have agreed to support communications protocols with this error-detecting capability, they can be assured data received will be identical to that sent. When companies choose to use an EDI VAN as a communications intermediary, the VAN also supports error-detecting protocols. What this means to you, the business user, is that when an EDI data stream appears to have been communicated successfully, it has, in fact, been received completely and identically to that transmitted.

How Do We Know that Our Trading Partner Received and Accepted Our Information? Once the data have been successfully received, the receiving EDI translator evaluates the transmission for completeness and adherence to standard syntax rules and generates a functional acknowledgment that is transmitted back to the original sender of the data. As described earlier in this chapter, the standard contains a series of control envelopes that allow the receiver to determine if a complete transmission has been received. Each envelope is comprised of a header segment preceding the data to which it pertains and a trailer segment following it. To verify completeness, the receiving translator determines that it has received matching control numbers in associated headers and trailers, signifying a valid beginning and end to each unit. It then counts the units contained in the transmission and compares its count against the count in the envelope trailer. If they are the same, the complete unit has been received.

Once the translator has determined that it has received all that was sent, it examines the data stream for adherence to standard syntax rules, i.e., that it contains all mandatory segments and data elements; that the data contained in each field are the correct data type, numeric or alpha; and that all conditional requirements have been fulfilled.

The receiving translator reports the results of its examination of the incoming EDI data to generate a functional acknowledgment (FA) which is subsequently returned via EDI to the sending trading partner. The FA contains documentation of any errors found as well as notification of acceptance or rejection at the preagreed level: functional group, transaction set, or segment/data element level. For example, the trading partners may agree that an incorrect or incomplete transaction set will be accepted and acted upon even when it is received with errors. Or they may agree that any incorrect transaction set will be rejected and retransmitted. Finally, they may agree that if a transaction set is found to contain errors, the full functional group in which it was transmitted will be rejected. It is always the case that rejected data must be corrected and resent.

How Will We Know if Our Trading Partner Will Act upon Our Transaction Set? What this question is really asking is, "Will the receiver's application accept and process our transaction?" This is a business issue, not really an EDI issue, because the business application determines if it can act on the incoming transaction after the data have been successfully transmitted and accepted by the translator. For example, based on the translator finding a string of characters present in the mandatory product code field of the purchase order line item, it generates a successful FA. However, if the order-entry application determines that it is not a *valid* product code or if the order-processing application determines that the inventory is unavailable, it may not be able to be processed. It is always the business application that conveys the intent to act upon the transaction to the sender. For example, a purchase order acknowledgment transaction conveys successful processing of an EDI purchase order. However, an invoice or advance shipping notification transaction may serve the same purpose. Likewise the response to a successfully received and processed invoice may very well be a payment.

All in all, the ability to track EDI data far surpasses that of paper documents. When business people are given access to the ongoing status of electronically transmitted business transactions, they become much more comfortable with EDI and become proponents of expanding the use of EDI.

Data Security

Data security is always of prime concern, both intracompany and intercompany. For some transactions, security is of even more concern than for others. Two cases will be discussed here.

1. *Authentication.* Guarantees that a receiver can detect any variances between what was sent in the EDI data stream and what was received

2. *Encryption.* Makes the data stream indecipherable during transmission

Authentication. When the prime concern is to guarantee that the receiver can detect tampering with the EDI data, a technology called authentication is used. Here's how it works. Immediately before transmitting an EDI data stream, the sender runs a special program initiated with a predefined and secured authentication key. Using this key, the authentication program performs multiple calculations on the data stream. The result of these calculations is an eight-digit field called the Macro Authentication Code (MAC) that is appended to the original data and is transmitted along with them.

Upon receiving the full EDI data stream including the MAC, the receiver invokes its authentication program which performs the same computations as on the sending side using the same authentication key. If the data received are identical to that sent, the same MAC will be computed at the receiving site. Any difference in MAC shows that the data have been tampered with. Authentication is often required of a company asking its bank to transfer funds electronically via an EDI payment remittance transaction, to allow its bank to know that the requested transfer of funds is identical to that transmitted by the sender.

You can see that the security of this technology is dependent on the security of the authentication key at either end of the transmission. Ideally, the responsibility for maintaining each key is divided, with at least two people knowing only a portion of the key at each end and the key being updated regularly. As any key is valid only between a specified trading partnership, the business of key management, i.e., the generation and transmission of authentication key parts to the responsible parties, will be a very important component of this security system.

Because authentication is a very cumbersome procedure, it will probably not gain support in corporations unless it is a requirement of the receiving trading partner. The financial community and the government have intimated that authentication may become a requirement in the future. However, currently, there are few implementations using authentication.

Encryption. It is interesting to note that, while authentication provides a means to detect tampering, it does not camouflage the data in any way. In order to make transmitted data indecipherable to casual inquiry, data must be scrambled. This scrambling technique is called encryption. Again, the sender employs a key to initiate a program that encrypts (scrambles) the data. The receiver employs the same key to decrypt the data. Encrypted data are typically 1½ to 2 times as long as the original data stream from which it was generated. Because of the additional length, encryption also is not used unless it is required. The government has some applications for which it is requesting encryption, such as the scrambling of orders from which confidential information can be deduced. For example, if an outside party was to track the ordering of food for specific army bases, it could anticipate troop movements that would otherwise be secret.

To ensure total security, i.e., encryption and authentication, data must first be authenticated and then the full stream excluding the outside envelope, which identifies the sender and receiver, must be encrypted.

Here is how authentication and encryption are handled by a VAN acting as an intermediary between trading partners. With authentication, there is no impact on the VAN, who just passes the data along as usual. However, encrypted data may cause some procedural changes in normal VAN processing. Usually, at least some cursory validation is performed when a VAN receives EDI data. However, since encrypted data would be indecipherable, validation would not be possible. Also, if the VAN's normal billing algorithm is based on the number of EDI segments handled, segments would be impossible to discern in the scrambled data stream. The VAN would conduct some alternate form of counting, such as assigning a predefined number of characters to a segment. If billing is based on number of characters, there is no problem. However, the bill would be substantially higher because of the increased length of the scrambled file.

Transmitting Data Directly or through a Third Party

Another issue that must be addressed when implementing EDI is whether to transmit data directly between trading partners or to use an EDI VAN as a communications intermediary. This decision, which has both business and technical implications, will directly influence the amount of in-house support EDI will require during the implementation period, and ongoing as well.

Figure 5-4 illustrates how EDI is handled directly between three trading partners. For each trading partner, a communications configuration and an *operational window* (hours when transmissions can be sent) must be agreed upon. In addition, the company initiating the communications session must separate its EDI stream into separate files, one for each partner, and transmit the appropriate data directly to each one. If the company sends data to and receives data from each trading partner only once each day, it needs to handle between 6 and 12 transmissions daily. For each trading partner, the company would need to support a different

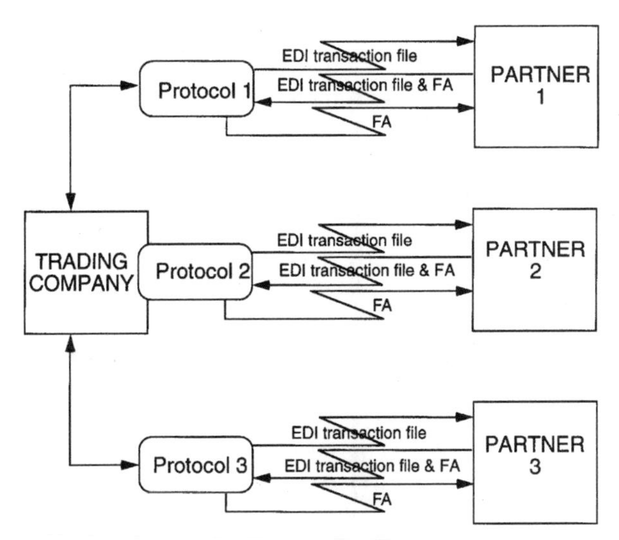

Figure 5-4. Direct EDI communication between trading partners.

protocol and/or line speed. It would initiate a communication session and send that partner's EDI data, receive both EDI transaction sets and a functional acknowledgment responding to its previous transmission of data, and return a functional acknowledgment for the EDI transaction sets received from the trading partner. The company may eliminate one communications session by holding today's functional acknowledgment until tomorrow's transmission of EDI transaction sets. Usually return of a functional acknowledgment within 24 hours of receipt of data is agreeable to an EDI sender.

To complicate the issue, communications sessions sometimes do not go to completion or a busy signal is received and the communication must be retried. Also, sessions need to be scheduled during the operational window during which the recipient is prepared to receive them. With up to 12 transmissions daily for only three trading partners, direct communications has quite an overhead of equipment and personnel. Not only that, but when a company chooses to handle EDI transmissions directly, it must be prepared to receive data from the VAN as soon as they are received from its partner.

On the other hand, Fig. 5-5 illustrates EDI through an EDI VAN. Here, the sending company prepares an EDI transmission file with data to be routed to all three trading partners and transmits the data in one session to the VAN. During the same communications session companies retrieve from their mailbox all data and functional acknowledgments which had previously been routed to their mailbox, as well as all system

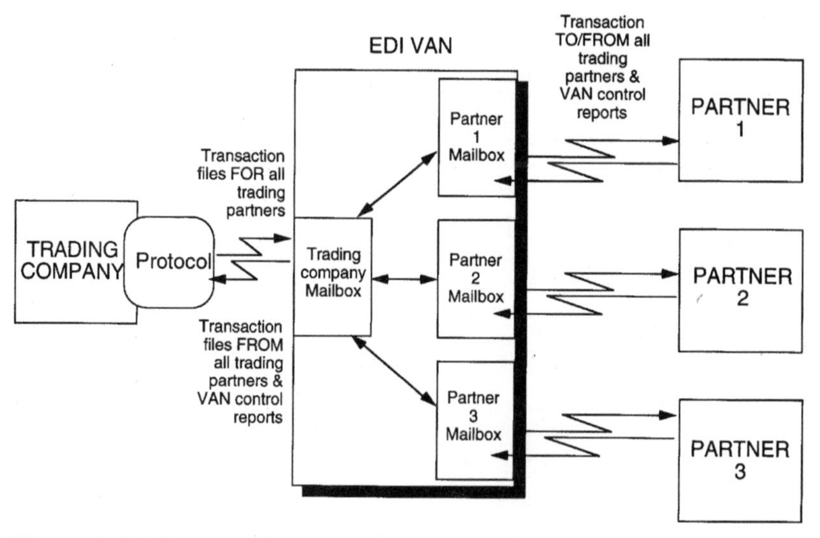

Figure 5-5. Electronic data interchange through a VAN.

reports generated since their last mailbox pull. Not only can all their EDI activity be limited to one or two communications sessions daily, but they can use one communication configuration (protocol and line speed) for all their EDI activity and can transmit and receive at a time of their own convenience regardless of when their trading partners choose to send or retrieve. Each trading partner realizes a similar convenience when dealing with a VAN. Finally, when a company supports direct communications, it must develop a system to track its ingoing and outgoing EDI transmissions.

The tradeoff for all this convenience and control is, of course, VAN charges for its various EDI services. As you implement EDI, it would certainly pay to compare VAN charges to the cost and impact of EDI on your operations and user departments. To assist you in making this comparison, a discussion of the various services of an EDI third-party VAN follows.

Services of an EDI VAN. A VAN provides several important services to both senders and receivers of EDI data. (Fig. 5-6.) In its role of communications intermediary, it allows a company to trade electronically with many trading partners while supporting only one communications protocol and line speed configuration. In this role it assigns each customer an electronic mailbox into which it deposits all the data and control reports destined for it. Control reports track the flow of EDI data to and from customer mailboxes and contain the date and time when data are received by the VAN and when they are retrieved by the trad-

Basic services

- Performs protocol and line speed matching
- Provides electronic mailboxes and EDI control reports
- Supports communications for all trading partners

Additional services

- In-network translation
- Links to intracompany E mail products
- Offers communications protocol and translation software products
- Offers a variety of routing options
- Conducts trading partner marketing programs

Figure 5-6. Services of an EDI VAN.

ing partner. These reports can and should be used by each VAN customer to verify that transmissions have been received in full and to develop an audit log of EDI activity.

The service of linking senders and receivers as well as backup and recovery of files on request are considered basic EDI services. While all VANs offer them, it pays to investigate their offerings and prices to determine which best fits your needs.

In addition to basic services, several VANs offer more extensive services such as:

- In-network translation
- Links to intracompany messaging products such as E mail
- Communications protocol and translation software products
- A variety of routing options
- Cosponsored trading partner marketing programs

Each is worth discussing as it may prove useful to you in developing your EDI implementation plans.

In-Network Translation. As was mentioned earlier, application programs never generate or accept EDI standard files. So, every EDI-active company needs to perform translation to and from the EDI standard. Translation can either be performed in-house by translation software, or it can be performed in-network (i.e., by an EDI VAN). If translation is performed in-network, the usual scenario is for the VAN to predefine the fixed-length file that it will translate to or from the standard format. When a company opts to purchase in-network translation services, it need only develop the application link that either generates the fixed-length file from data available out of its business application, or accept and pass the fixed-length file to its internal business application. In effect, in-network translation insulates the company from the EDI standard by eliminating the need for company personnel to know and work with the standard.

Sometimes, a VAN will allow its customer to predefine the fixed-length file, enabling transmission of that file directly from the customer's computer application to the VAN's translator. This eliminates the need to develop the application link module at all.

Another use of in-network translation is to eliminate the need to support more than one standard. Take the case of a company that has trading partners in two industries, each of which has developed an EDI standard. For example, a drug wholesaler buys health and beauty aids from manufacturers who are most likely supporting the UCS standard and pharmaceuticals from drug manufacturers who are probably supporting the ORDERNET/NWDA standard. The wholesaler examines its trading partner base and chooses to support the standard that is supported by the majority of its major trading partners. In this case let's suppose that standard is ORDERNET/NWDA. In order to deal with its health and beauty aid suppliers it would have to either develop a second in-house translation capability or purchase in-network translation services. In fact, it may be cost-effective to do the latter.

As EDI activity continues to grow, the use of in-network translation continues to dwindle for two reasons. One, the overwhelming majority of EDI-active companies purchase a translator of their own. Most translator products support all the variable-length EDI standards, including UCS, TDCC, ANSI X12 and all its subsets, and EDIFACT. Two, the number of companies supporting fixed-length industry-specific standards such as ORDERNET/NWDA or EAGLE, not able to be translated by a purchased product, also continues to dwindle as the companies convert to one of the variable-length standards.

Links to Intracompany Messaging Products. Electronic data interchange is the intercompany component of messaging between trading partners. As EDI activity increases, there is a growing need to disseminate information received via EDI messages to various parts of the company. While E mail has traditionally been used only for sharing of intracompany information, it is beginning to be employed also as a precursor to EDI. In this role, it collects and passes information from one site to another to develop a complete transaction. Although E mail is best known for free-form, print messages, in its role of information collector, it is more useful to have predefined forms into which staff members can enter data. By structuring E mail forms, information in the form can be interpreted and mapped from the screen fields to a standard X12 transaction set.

Electronic mail can also be used on the receiving side by moving information from an EDI transaction set into predefined E mail messages and then distributing these messages to internal business and technical users within the organization. To provide the most useful information to the business and technical community, potential receivers of E mail messages should define the data fields they want and need to see. The technical support group would use these requirements to design the E mail message that would be distributed in-house. Some EDI VANs offer E mail products and other electronic commerce messaging and information-sharing products such as UPC catalogs.

Sale of Communications Protocol and Translation Software Products. In order to fulfill your communications and translation requirements for EDI, you may need to purchase communications software and will probably need to purchase an EDI translator. While these products may be purchased separately from various sources, you will find that it will greatly facilitate your implementation to purchase them from a single source.

In this way you will not only purchase products which fulfill your processing and communication needs but will be assured that all components are compatible with one another and with the third-party VAN.

Variety of Routing Options and Enhanced Services. In general, EDI is used as a store and forward capability. The information received by an EDI VAN from a sender is stored in an electronic mailbox awaiting the receiver's retrieval request. It is then forwarded to the receiver. This scenario requires that the receiver inquire regularly for messages, which is not always an efficient retrieval method. As an alternative, some VANS offer additional routing options. Two such options are *timed-outdial* and *event-driven delivery.* In the former, an EDI receiver who is a VAN customer requests that the VAN automatically send the data to them at specified times per day. While this does not provide instant delivery in case of an urgent message, it guarantees that no data will remain in a mailbox for more than a predetermined number of hours. Of course, this means that the receiving company's communications capability must be available for receipt of data at the preagreed transmission times. For the latter, event-driven delivery, the receiving company defines the event, and the VAN automatically outdials and transmits data to it when the event occurs. Some of the most commonly used event triggers are:

- Volume of data
- Time period
- Trading partner identifier
- Transaction set type

For *volume of data,* the receiving company predetermines the maximum number of interchanges it will allow to be stored in its mailbox. When this number is reached, the VAN automatically transmits the stored interchanges to the company. This is used to help the receiver to load-balance its daily processing of EDI transactions. For *time period,* the receiving company specifies the maximum number of hours that it wants data to await retrieval before triggering an automatic outdial to the company. For *trading partner identifier,* the receiving company identifies for the VAN the trading partners from which it wants to receive transactions immediately. When data are received from one of these partners, the EDI VAN immediately outdials the data to the receiver, who presumably is prepared to act on them immediately. For *transaction set type,* the receiving company specifies for the VAN which types of transaction sets it wants to receive immediately. For example, it might want purchase orders or purchase order changes immediately but will allow invoices to be stored awaiting retrieval.

By defining the event triggers that best support your business needs, you can increase your own productivity and be more responsive to trading partners at the same time. Vendors who offer vendor-managed inventory programs, discussed in Chap. 8, use these delivery options as an integral part of their programs. Customers receiving advance shipment notifications from their vendors use appropriate delivery triggers to assure that they receive their transactions quickly and have time to prepare for timely product delivery.

In addition to the delivery options above, which deal with the speed at which EDI data is received, there are other options that deal with the mode in which it is delivered. Three such modes that are offered by EDI VANs are:

- Carbon copy
- Fax
- Express and electronic mail

Carbon copy is used to provide a duplicate copy of an EDI transaction to an additional recipient. Either sender or receiver may request that one or more additional parties receive a copy of the EDI transaction set. This triggers the network to automatically generate copies and distribute them to the companies named by the carbon copy requester. There are various uses for this service. If you are a customer buying directly from a manufacturer but expecting delivery from a wholesaler-distributor, you will want a copy of your purchase order sent to both parties, one to transact the purchase, the other to alert them of shipping requirements. Or if you are a vendor, sending an invoice to your customer's corporate accounts payable department, you may request that a duplicate invoice be sent to the ordering location as well so the customer may verify that product has been received and authorize payment of the bill. This is a very useful option. You can probably think of many additional uses of it that can help to streamline your business procedures.

With the *fax* and *express* or *electronic mail* options, the VAN converts the EDI transaction set to a print format before passing it on to the receiver. For the fax option, the print format is transmitted via facsimile machine; for express mail, it is sent via mail service; and for electronic mail, it is transmitted to the receiver's E mail system.

Although these options do not send machine-readable data to the receiver, they are particularly useful when trading with partners who are not capable of receiving EDI data, especially when a company attempts to convert all of a particular type of business to EDI. This allows a company to reach those trading partners who show no signs of ever becoming EDI-capable but with whom it intends to continue to trade.

There is one important limitation to this capability. While it is fairly easy to convert from machine-readable EDI to human-readable print, it is almost impossible to convert going from print to EDI. So, this solution is virtually a one-way street. The responding trading partner will not even have the capability to return a functional acknowledgment to report on successfully receiving messages. Luckily, it is rarely used between partners that conduct substantial amounts of business with one another or even between those who trade regularly but in small quantities. For those trading partners with whom a company trades sporadically or infrequently, this solution is preferable to continued support of their paper system.

Trading Partner Marketing Programs

Since EDI must be performed with trading partners and EDI benefits are realized in direct proportion to the amount of business transacted electronically, in effect, the more partners, the more benefits.

Often, a great deal of time and effort is needed to either encourage partners to implement EDI or to provide guidance for and testing with each one. Some companies have dedicated staff to bringing up EDI partners. They not only sell the concept of EDI, but provide support during the implementation and testing phases. Other companies either cannot afford to take this route or choose not to. Unless they happen to be a company with enough clout to pressure partners to do EDI, they have a difficult time "growing their program." Even when a company devotes staff, they find that after the staff has brought up the company's experienced and most enthusiastic partners, the marginal amount of work needed to bring up additional partners increases. Often the project falls into a somewhat static state with little or no growth over a long period of time. When this happens, the lack of success in this EDI application may act as a barrier to future EDI implementations in other divisions or business areas within the company.

In response to this problem, EDI VANs willingly offer their support to help bring up EDI trading partners. Several have programs through which they proactively market for and support the implementation and testing of EDI by their customers' trading partners. When you select a VAN, be sure to ask if it has such a program. Ask for references from other companies that have used its marketing and trading partner support programs. Be sure to check with these references to ascertain how successful the VAN was and what type of response time was provided from the time the reference company gave the VAN its trading partners' names until they were contacted by a VAN representative. Also ask about the timeliness and accuracy of VAN reports for tracking the status of prospective EDI partners.

We will make just two final notes on EDI VANs.

1. A trading partner marketing program is only successful when it is a joint venture between yourself and the EDI VAN, where you, as the EDI driver, provide the business impetus, asking your trading partners to do EDI with you and outlining the potential benefits of converting from paper documents to EDI. The VAN provides the marketing staff to facilitate the trading partners' implementations. This type of cooperative marketing program is a win-win-win situation, offering benefits to all parties. The EDI driver brings up its EDI partners faster and with less support personnel, the VAN increases its EDI traffic and sale of EDI software, and the trading partner gets the advice and support it needs to implement EDI quickly and cost effectively.

2. While you will want to encourage your trading partners to do EDI with you, you will probably not want to dictate which VAN they use. When selecting a VAN yourself, be sure to look for one that regularly interconnects with other EDI VANs and has reasonable interconnect

charges. Today most VANs interconnect with each other. However, they differ in the speed at which they pass data and in the type of controls and reports they have to track internetwork communications. As a prospective customer, you will want to look for the most cost-effective and convenient service with the strongest customer service and support organizations.

EDI and International Trade

Using EDI to replace the vast amount of paperwork required for international trade has proven to be an excellent application. As in the domestic arena, standards development has been the key to successful implementation of international EDI. Prior to the development of EDIFACT, the internationally supported EDI standard, it was virtually impossible to trade electronically using only the existing industry-specific EDI standards. However, with EDIFACT so closely related to X12 and having the same architecture and syntax rules, international trade via EDI has become fairly commonplace and easy to support.

As you can imagine, there are some differences between the North American standards and EDIFACT. Segment names, their order within a transaction set, and field names and placement within the segments vary. Also, usage of segments and fields may vary. For example, a field that is mandatory in the X12 standard may be optional in the EDIFACT standard. A segment that is optional in North America may be mandatory in the international standard. Additionally, the European method of field usage has been retained in cases where North American methodology was unable to provide the functionality required. For example, ANSI X12 names and addresses are highly structured into four separate address lines—the first for name, the second for additional names, the third for street address, and the fourth for city, state, country, and postal zip code. That works perfectly well for North Americans, as our addresses have a standard shape. However, internationally, addresses take various forms. Therefore, the international standard has a more flexible structure containing one long field for the full name and address. Data for each part of the address are separated by a control character known as a *subelement delimiter.* The receiver reads and interprets the name and address by separating delimited fields.

Even though there are small differences between North American and EDIFACT standards, the information required to transact business internationally is almost identical to that needed domestically.

You are probably wondering if the international standard can be handled by your domestic EDI software. The answer is most probably yes. Most translators can handle all variable-length public standards of which EDIFACT is one. However, there are some additional tasks required to successfully bring up an international partner. Assume you want to send an international invoice and that you already produce a domestic X12 invoice. To add an international trading partner, start with the same up-front work as to add a domestic partner:

- Enter company and communications information from this partner into your EDI translator's trading partner profile.
- Add this partner's transaction set requirements to your EDI translator files.

Then, if there are additional information requirements, upgrade your system to provide the additional fields in the outgoing EDIFACT message. The same steps are required when you are the receiver of the international transaction; however, instead of generating it, you will need to validate and interpret incoming data. When mapping the fields to your internal format, you may find that you need to enhance your current business application to accept additional pieces of information not included in the X12 standard. For example, you will need to prepare to receive currency identification as part of an EDIFACT transaction set, although you rarely get it in an X12 set.

Several EDIFACT transaction sets have already been developed. The overwhelming majority of them are used for transportation and billing.

Payment Terms in the EDI Environment

Many years ago, payment terms were offered to a payer by a supplier as an inducement to get paid sooner. Terms were arrived at by computing the opportunity cost of capital. Historically, terms have remained static from that time until the present, while prices, interest rates, and the like have fluctuated up and down.

Let's take a look at a common terms discount of 2% 10 net 30, meaning take a 2 percent discount if you pay within 10 days of invoice date; full payment is due within 30 days. When the invoice was received 3 to 5 days after it was generated and mailed, returning payment within 10 days was not easy. The inducement, of course, was the 2 percent discount. However, when payment was mailed within 10 days, it was not received, deposited, and available for use for many more days. With EDI the invoice can be received within minutes or hours. This has caused companies to rethink their terms arrangements. Since payment is triggered by payment terms, it would seem that there would be mutual benefit derived by renegotiating terms for the electronic environment. Consequently, buyers and sellers often use the implementation of EDI as an opportunity to reopen the terms discussion and to reevaluate the opportunity cost of capital in the current business environment. They then arrive at new terms that are mutually beneficial to both parties. Sometimes they agree to increase the discount period from 10 days to 12 or 13 and continue to offer the same discount; sometimes they increase the discount and leave the period the same.

Not only is the invoice received sooner via EDI, but reengineering of business procedures and development of automated reconciliation systems allow companies to compare the incoming invoice with the open purchase order and receiving information and to authorize payments much sooner as well. With all this accomplished in a matter of hours instead of days or weeks, payment can be transacted much faster. In the past, eliminating float through EFT was not popular among payers. However, with new banking regulations all but eliminating interbank settlement delays, EDI is gaining support as a cash management tool.

Business Arrangements in the EDI Environment

It is important to remember that merely imposing EDI on current business procedures will not deliver long-range savings. Yes, you will spend less on paper and envelopes and you will certainly save mail time, but in order to reduce inventory and eliminate long and costly reconciliations, business procedures must change. However, if they change in a vacuum, i.e., within your company but without impacting intercompany procedures, they will have little long-term effect on improving business relationships with trading partners.

Timing

Because speed and accuracy are keys to good business relationships, both buyers and sellers seek out opportunities to share accurate information in a more timely manner. Some examples: A vendor shares its order-processing schedule with customers to allow them to synchronize their own purchasing to the earliest processing time. A customer shares forecast information to allow its vendor the opportunity to manufacture for the expected demand and to forward-buy from its own vendors at favorable prices.

Additionally, EDI-active companies share information regarding their operational windows, those hours per day and days per week during which they can receive transmissions. Here timing is the difference between successful and unsuccessful transmissions. Two companies communicating directly from opposite coasts will find few hours in a normal workday when both are open and able to communicate. They must preagree on specific transmission times so they can successfully trade electronically.

Not only is timing important for the initial processing of a transaction, it is equally important for acknowledgments. There are two standard mechanisms for acknowledgment. The first, a transmission acknowledgment (TA), is rarely used. The TA is generated by the receiving translator even before syntax is checked on the incoming data. It merely warrants that the transmission has been received, but makes no value judgments on the data. The second, the functional acknowledgment (FA), is also generated by the receiving translator. It also warrants receipt of the data. In addition, it reports on completeness and correctness of the data stream. Acceptable time frames for acknowledgment are agreed to in advance by trading partners. The question is often asked, "What is an acceptable time frame for acknowledging receipt of a transmission?" The answer is, "As quickly as possible." Most companies translate incoming EDI data as soon as they receive them, and their translator automatically validates the incoming data stream and produces an FA to be returned to the sending partner. The FA is ready within minutes of receipt of data. The final step is, of course, to return the acknowledgment to the sending partner. Often this step is delayed until the next day's communications session to the EDI VAN.

Usually, 24 to 48 hours is a practical time period for receipt of either the TA or FA. Once the partners have agreed upon an acceptable time period for return of an acknowledgment, each needs to track his or her own activity. The receiver of EDI data generates and returns the acknowledgment in a timely manner, and the sender retrieves it, reconciles it to his or her EDI data counts, and follows up on exceptions.

To further improve timeliness, many companies automate the processing of acknowledgments for tracking EDI data as follows. The sender's system peruses the dates and times when the previously transmitted EDI files were transmitted. It identifies those for which no acknowledgment has yet been received. By comparing the elapsed time since transmission of the EDI data with the agreed upon period of time in which an acknowledgment was expected, the tracking system alerts the appropriate people of potential problems. The people then follow up with their trading partners. When an acknowledgment is received within the agreed upon number of hours, and when the counts and transaction set identifiers agree with those saved by the translator, no alert nor intervention is necessary.

Another instance when timing is important is in handling and processing of change transactions. Electronic transmission provides rapid intercompany movement of data. However, processes such as scheduling a manufacturing procedure, tooling for it, and performing it, take finite amounts of time and require lead time. Receiving a change transaction after a procedure is set up or has begun can cause not only inconvenience but loss of time and money as well. For this reason, companies must preagree on the lead time required to handle EDI change transactions and the procedure to follow when less than the agreed upon lead time is available. Most companies choose to handle change transactions off-line by telephone or fax. Even companies that have received EDI purchase orders for many years often continue to process purchase order changes manually.

Summary

In this chapter we have dealt with the most important business issues you will need to confront when implementing EDI in your organization. We have seen that EDI impacts all aspects of current business and system environments as well as intercompany relationships and procedures. On the business side, both procedures and staffing requirements change. On the technical side, new systems are often needed, and existing systems must be enhanced to fulfill new requirements such as:

- Accepting data files instead of the traditional key entry
- Providing and accepting additional information to fulfill EDI standard requirements
- Performing newly automated processes and decision-making routines
- Keeping track of EDI activity in and out of the company

Without the existence of an acceptable standard, EDI would never have grown as it has. However, more than one standard has been developed. Some support the needs of a particular company or industry group. One, ANSI X12 supports cross-industry application. EDIFACT supports international trade. In selecting an EDI standard, a company must evaluate its own data needs and those of the industry in which it does the majority of its business. Most often, the standard already being supported by its trading partners is the one the company decides to support itself. Companies trading in multiple industries may need to support more than one standard.

In addition, there are several security issues to be addressed for the electronic environment:

- Replacing paper (hard) copy with machine-readable data files
- Eliminating the authorizing signature
- Ensuring that electronically transmitted data sent were received in full
- Protecting data from casual inquiry or change during transmission
- Developing a complete audit log of activity

It is important to address these issues from the outset of the EDI project and to include control and security measures in your plans for implementation. Unless the business community has confidence in the EDI system, it may work but will be a little-used system. Electronic data interchange actually gives us the opportunity to improve controls and security by tracking and reconciling the accurate and timely information sent and received.

Most importantly, EDI works best with those trading partners with whom you have an excellent working relationship. Unlike paper documents in which most processing is manual, electronic trade processes machine-readable data in an automated way, without human intervention. Surprises are impossible to handle in this environment. Be sure to work out data requirements, processing time schedules, and change transaction processing prior to implementing EDI. You may also want to take this opportunity to reevaluate your payment terms and discounts. Often, both parties can benefit from a renegotiation of traditional terms.

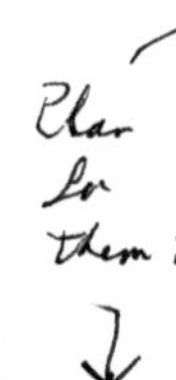

As you have seen in this chapter, the lion's share of the effort involved in evaluating EDI opportunities and planning for its introduction is business-related. Later on in the project, the largest share of implementation and support activities is technical. Most companies have found that a successful EDI implementation requires a partnership of business and technical groups working together.

6
The EDI Teams

Objectives

- Identify the three teams needed for a successful EDI implementation and the ongoing support of an EDI program.
- Identify the members of each team and describe their responsibilities.

Building the Right EDI Teams

When an EDI program takes the proactive path outlined in Chap. 5, each new application of EDI goes through three distinct stages. It starts with a planning stage, then goes through a formal implementation period, and finally proceeds into an extended ongoing growth period. During each of these three phases, there is a different team of support people needed to perform various tasks. For a reactive EDI program, implemented in response to pressure from a major partner, there are often only two stages, implementation and then growth. The planning stage is often bypassed. As you'll see, this serves to not only limit the scope of the initial use of EDI but also acts as a deterrent to future uses.

The typical entry of EDI into an organization is at the middle-management or lower level. For the company driving the EDI implementation, usually the customer, this is because inefficiencies have the most impact on the manager or business user in a functional area. For example, it is the purchasing manager who is under pressure to get the right products on the shelves or available to the manufacturing process at the right times, while keeping inventory expense to a minimum. It is the accounts payable manager who must handle and reconcile incoming invoices with a minimum of error and with limited resources. On the vendor side, EDI is implemented in direct response to trading partner requests. The typical scenario is for the requester's purchasing, or procurement person to approach the trading partner's sales and customer service person and ask him or her to accept orders electronically. Based on the traditional segregation of job functions in a company, the salesperson faced with this request can only influence system changes within the sales department. The salesperson can do little toward a total overhaul of internal procedures and organizational structure. So, wanting to please the customer, salespersons will focus on installing the technology of EDI and charge the IS department with the task of implementing it. Unless upper management can be brought into the discussion to authorize a more strategic approach, the resulting EDI program will be very limited in scope. It will rarely impact more than the systems of this one department, often not even automating its business procedures. For the vast majority of EDI implementations, this is the case. In the short term it seems like a reasonable approach. Not only that, but with time pressures imposed by the trading partner, it may be the only practical approach. *"If the customer wants us to accept EDI orders, that's what we'll do!"*

The problem is that when EDI is implemented with only the trading partner's focus, it is often difficult to get out of the reactive mode and short-term approach. Few companies step back after this first application and develop a plan for implementing EDI in other parts of their companies.

The bottom line is that even when EDI is first implemented in response to a trading partner's request, it should still be thought of as a strategic initiative, and its broad and long-range possibilities should be investigated. So, whether planning and evaluation precedes implementation or follows it, they are both essential in the development of an EDI program that directly responds to your own business needs.

Let's discuss each of the three teams referred to above and how they support the EDI project during its various stages.

The EDI Steering Committee

This committee's main function is to set the corporate strategic direction of EDI. It is active during the planning stage of each new EDI initiative. As you can imagine, the members of this committee should be those who have the authority to set and change strategic direction. While specific job titles may vary from one organization to another, it is usually the case that higher-level people are appropriate candidates for membership. When this committee acts as a high-level advisory board, reporting directly to the president or chief executive officer of the company, there is high likelihood that the EDI direction and goals that are approved will be pursued more aggressively and will be attained. However, when the committee members are midlevel managers, reporting too low to gain the commitment of top-level management, they rarely muster the clout to get the right people involved or to make any substantive changes. While it is typically *not* the case that the EDI steering committee sets a predefined corporate direction for EDI, the members often discuss how EDI and other related technologies of electronic commerce can be used to support corporate goals and objectives and augment current and future plans. In fact, based on the company's goals and directions, they will select different applications of EDI and related technologies. For example, a company whose goal is to increase market share at the expense of major competitors will look for ways to outperform its competitors. This company will look for opportunities to offer new high-tech and other value-added services attractive to customers which will encourage them to move their business. In order to do this, the company may decide to implement various EDI transactions with customers, which will allow it to be more responsive to customer needs. The company may also identify new and innovative ways in which it can use the more timely and accurate EDI information.

On the other hand, a company facing fierce downward price pressure from its competitors will focus on cost-cutting measures to maintain its profit margin. For this approach, it will try to reduce its inventory levels and streamline internal procedures. Its EDI direction would very likely be to initiate EDI with its vendors, initially sending orders and then asking for invoices and shipping notifications in return.

In the very longest term, both of these companies would converge on the same or very similar electronic commerce programs, converting all their most commonly used business transactions with both vendors and customers to EDI. However, unless the short-term direction responds to a real business need and brings the company closer to its corporate goals, there is a very dubious prognosis for future EDI implementations.

Because of the breadth of EDI opportunities throughout the organization, it is best when this committee is comprised of members from all the functional areas. It is also helpful to have a person who is already knowledgeable about EDI at committee meetings to describe how EDI is being used to best advantage by others in this and other related industries. This is usually a company person who has been charged with learning about EDI and investigating its opportunities for just this purpose. Often this person is called an EDI coordinator. We'll be talking more about the EDI coordinator later in this chapter. Finally, this committee should have a high-placed IT person as a member whose role is to consider the use of technology to support the various corporate initiatives discussed.

- High-level representatives of the various functional business areas
- High-level representatives of the information technology department
- EDI coordinator
- Other corporate executives involved in developing strategic objectives

Figure 6-1. Steering committee members.

When members of this committee determine that EDI can actively further corporate objectives and are of sufficient level to carry out their intentions, they make it happen by expressing their strong commitment and approving the use of the people and dollar resources needed. While this committee is needed to set strategic direction, it must pass the baton to the functional areas and IT support staff to determine how to proceed.

The EDI Implementation Team

The EDI implementation team is responsible for the actual implementation of EDI. The deliverables from this group are:

- Business feasibility study
- Technical evaluation and recommendation
- Communications evaluation and recommendation
- Legal recommendations and EDI trading partner agreement
- Auditing requirements
- Employee training plan

In short, this group analyzes the "why to" developed by the steering committee and develops the "what to do" for the company's business areas and their systems. Further deliverables from this group are:

- Business implementation plan
- Prioritized list of targeted partners
- Technical implementation plan
- Specifications for system enhancements
- EDI translation product recommendations
- Communications options and recommendations

- Managers of the various functional areas
- System analysts from the information technology department
- EDI coordinator
- Representative from the legal department
- Representative from the auditing department
- Representative from human resources

Figure 6-2. Implementation team members.

In order to accomplish all this, the team must also have active participation of the people who can make it happen. In successful EDI programs, the implementation team is often comprised of:

- Business managers in charge of the functional areas into which EDI is being implemented
- IT personnel who support functional area computer applications
- Communications specialists
- Legal and audit representatives
- Human resources representative

Each person has a specific role to fill. In small companies, one person often fills more than one role. For example the IT representative and the communications specialist are typically one and the same person. At the least, there must be both business and technical representation on this team.

Functional Business Tasks. On the business side, functional managers determine if and how EDI can be used to help the functional area to become more streamlined and efficient in conducting its own business and in furthering corporate objectives. As part of the business person's participation in this effort, he or she can gain their department's assistance in producing the functional area feasibility study which will be described in Chap. 7. Once the feasibility study is completed and evaluated for potential savings, it provides the business case for EDI. Business requirements defined in the feasibility study are passed to the technical IS and communications members who determine if and how they can be incorporated into the current system and communications environment. Often they recommend enhancements to current business applications. Sometimes, they will recommend major rewrites or development of new systems.

Because functional business managers have first-hand knowledge of the companies with whom they do the most business and are already the natural liaisons with partners, they take this opportunity to select the partners they wish to target for EDI and to develop overall goals for each application of EDI. They may even decide to gain additional information about trading partners by surveying them regarding their EDI plans. From this additional information they develop a plan for phasing in EDI trading partners over time.

Technical Tasks. IS managers on the implementation team pass the business requirements defined in the feasibility study to a system analyst who evaluates them and identifies their impact on the systems environment. The technical people develop a technical plan that describes which systems need to be upgraded and which new systems are needed. Typically, when the business community identifies requirements for accessing information received via EDI, technical development work is required.

The technical people also use the business plan to determine what functionality is needed in an EDI translator. They then evaluate available products and develop a product purchase recommendation. Finally, the technical people examine business information requirements and determine how they will be represented in the EDI standard. Based on completion of the above, the technical people develop a plan to proceed as well as a schedule for completion of their tasks.

Legal and Audit Tasks. You may be asking why legal people and auditors get involved. Typically, they are asked to attend only occasionally during the implementation team's regular meeting schedule. Legal and audit representatives are asked to participate to answer such questions as

- Do we need to have a special EDI trading partner agreement in place to deal with a company electronically?
- If so, what should be covered in the agreement?
- What business and processing information do we need to track in the electronic environment to develop an audit trail of EDI and intracompany activities?
- What access to information does the auditing staff need?

By inviting legal and audit representatives to EDI implementation team meetings, we allow them to learn of our plans and to provide input as to how the above and other issues must be addressed. Instead of being outsiders to the EDI process, these members will become active participants. Auditors in particular become strong supporters and promoters of EDI because its machine-readable transactions have a higher degree of accuracy and its control aspects provide the ability to more accurately track activity than do paper counterparts.

Human Resources Tasks. Finally, here are several reasons why I suggest that the implementation team invite a representative from the human resources department. One, so that human resources understands what EDI is and where it is being introduced. Two, to explain to the department how EDI will impact the work environment and individual jobs within the functional areas. Three, to allow for the opportunity to arrange training for employees whose new job responsibilities will require it. And four, so human resources can prepare to address employees' concerns regarding change within the organization. Even when a change promises tremendous improvement, human nature is such that people are fearful of it. Handling change within the organization is one of the most important aspects of successfully introducing EDI and automation into an organization.

Full Team Tasks. This team is responsible for seeing through the EDI implementation from inception to completion. Completion in this context means that two activities have been successfully accomplished. One, your EDI system and automated business procedures are in place and working, and, two, at least one trading partner is up and running with you. As we have already discussed, bringing up one trading partner is rarely the ultimate goal of the program. So, as a team, this group will want to use its experience in bringing up this first partner to develop an EDI pilot plan that defines all the tasks needed to bring up each new EDI partner. This plan is passed to the EDI support group who use it as a template when bringing up additional partners, adding new tasks as needed.

Another team task that goes a long way toward influencing the attitude of executives toward EDI is reporting of the progress the company is making toward its EDI and electronic commerce goals. The team should provide reports on a regular schedule that stress the results it has achieved in terms of number of partners, percent of converted business, and savings realized. This will not only provide the latest EDI news but will serve to sell EDI to additional internal users as well.

The EDI Support Team

Once the implementation team has successfully completed and tested the intracompany part of the EDI project (application link, EDI translator, communications, business procedures), and once it has successfully brought up at least one partner, the project is typically turned over to a support team. The members of the support team perform the tasks needed to

- Bring up additional partners
- Resolve issues that arise
- Maintain the currency of the system

Typically, this team resides in the IT department and is comprised of IT staff who handle syntax-related issues as well as the logistics of actually transacting business electronically. In addition, the team contains a business user to handle content-related questions and issues that may arise. Because of the variety of tasks needed during this stage, and the different levels of expertise and areas of specialization required, there are often several people on the support team. In smaller companies or companies with minimal IT departments, one person may wear many hats and be responsible for several categories of tasks.

Bringing Up Additional Partners. If the pilot plan has been developed as described above, the support group performs each of the steps specified in the plan when bringing up each new EDI trading partner.

Handling Business Impact and Data Content. Either preceding this plan, or as step one, the functional business person works with each prospective EDI partner to discuss the business implications of converting business to electronic form and to develop an implementation schedule to proceed. As additional tasks with this partner are performed, the functional business person continues to be the main contact for content and business-related issues that arise. It is the functional business person's job to discuss with the partner the information that needs to be shared in an EDI transaction set and what will be done when an EDI transaction arrives with missing or unnecessary information. In general, issues involving the content of EDI transactions and the business relationship with partners are the province of functional business users. Issues involving syntax and operations of the system are the province of the IT members of the support group.

It is important that the various issues that arise be predefined so they can be detected automatically by the system and passed to the appropriate person for resolution. The more you automate the problem detection function, the more you reserve resources to resolve issues. Automatic detection can be implemented simply by programming various components of the EDI system to look for the conditions that you have defined as exceptions, alert the party responsible for working with this type of exception of its occurrence, and provide a user-friendly way of resolving the problem and reinserting the affected transaction into the normal processing stream.

Finally, the IT people on the team handle issues that arise in conjunction with business systems, the application link, the translator, communications, and the EDI VAN.

Working on Business Systems and the Application Link. Application programmers are assigned the responsibility of working on both the business system and application link. If new data fields are required by or received from the trading partner, the support team must provide a path for these fields to move between the system and application link programs. Most often, once testing has been completed with the first few EDI partners, system and link issues have been encountered and resolved. For an EDI sender, this support group adds each new EDI partner to a routing table that instructs the system to pass this partner's business transactions toward the EDI system instead of toward paper processing. For an EDI receiver, they must provide a partner-specific link. This requires collecting information from the EDI transaction set, passing it between the translator and business system, and performing special procedures for this partnership.

Setting up and Maintaining the Translator. Once additional partners begin to be brought up, various partner-specific issues arise. Often different data are required or will be received. Sometimes, different procedures are performed. Working with the translator is a major part of the ongoing support of EDI. Some translator tasks are performed as part of the pilot program for each new EDI partner, for example, adding the partner to the translator's trading partner profile by providing all the information needed to recognize the partner and process its data automatically.

Translator Mapping. In addition, the support group works with each new EDI partner to agree upon standard usage and then performs additional mapping, if necessary. The support member who works on this task should be knowledgeable about both the translator and EDI standards so he or she can set up new mapping and processing rules for the translator. If a new partner places no additional requirements nor processing on the translator, the support person merely adds this partner to the trading partner profile using existing mapping rules. However, when the new partner requires that you send additional or different data fields, the sender must instruct the translator to place this new information into the standard or to use different qualifying codes.

Setting up Communications. For each new EDI partner, the communications logistics and configuration (line speed and protocol) must be agreed upon. When a company chooses to communicate directly with each trading partner, this can be somewhat of a challenge. Not only do different companies support different line speeds and protocols, but they run their applications and transmit their EDI data at all hours of the night and day. For this reason, most companies employ EDI VANs. However, even when a VAN is used, there are still some tasks to be done for each new EDI partner such as

- Finding out which VAN the partner uses
- Making sure that your VAN interconnects with the partner's VAN
- Adding telephone number and communications configuration information to your EDI trading partner profile
- Testing and verifying that communications is occurring correctly

Monitoring Control Reports. Once a trading partner has been converted to EDI, ingoing and outgoing EDI data must be monitored for correctness and completeness. This is done automatically by the translator and results in control reports that note any errors encountered. Using these reports, support group members determine if data were correct and complete and processed in full. If not, they research the cause of each error and either pass it to the appropriate person or resolve it themselves. Most errors involving EDI data are encountered during the test phase with each partner. Once this phase is completed and the partner is moved to production mode, rarely are additional syntactical or content errors found.

Resolve Issues. The support team stands ready to resolve issues and answer questions as they arise. Once initial testing has been completed, ongoing problem resolution is usually fairly inconsequential. Occasionally exceptions are encountered within your own EDI system. For example, a transaction is processed by the business system but does not either get passed to or processed by the application link. When this happens, the support group must reinstate the transaction so it can be processed throughout the entire system and transmitted to the trading partner. Occasionally, an EDI VAN report shows that transmitted data were not successfully received and processed by the network or that data awaiting retrieval were not successfully retrieved by you. For these cases, the support group works with the EDI VAN to either send the EDI transactions again or to re-retrieve them.

Since the EDI support group is typically the main trading partner contact within your organization, it receives many calls from EDI partners. While some of these questions are EDI-related, it has become common practice that once business transactions travel electronically, all questions related to them become EDI questions. By using a simple example such as the following you can see how ridiculous this can be. When an EDI invoice is received, the customer calls the EDI support group to complain about its billing. This is akin to calling the post office to complain about the contents of a letter. However, ridiculous as it is, most support groups spend a good deal of their time trying to answer questions that they are totally unprepared to answer. The way to eliminate this time drain on the technical support group is to provide both business and technical contacts for your EDI trading partners and to clearly specify what types of issues are handled by each.

Maintain Currency of the EDI System. Because your company will bring up new EDI trading partners over time and because the standard undergoes changes over time, the support team must be prepared to install the newest version of the standard for new EDI partners as well as to maintain past versions for existing trading partners. In addition, EDI translators are updated on a regular schedule to incorporate extra functionality. This same group is responsible for installing each new version of translator software and verifying that it links successfully to the application link program and receives from or passes to the application all the required data fields.

Summary

Since the implementation of EDI and the installation of new EDI partners is an ongoing effort, it is important to have the right people involved at the right time. During the planning stage, this means having a steering committee comprised of strategic thinkers representing various parts of the company. The deliverable from this group is a plan of attack that supports corporate goals and objectives.

During the implementation stage, this means an implementation team that evaluates various EDI opportunities throughout the organization and prioritizes them based on their potential payoff. For each EDI application selected, this team performs the business and technical tasks required to set up the business procedures for electronic trade, set up the EDI system, and perform testing for the first one or few EDI trading partners. This team is comprised of functional business area representatives (usually managers), IT personnel, communications specialists, legal and audit representatives, and sometimes human resources as well.

Then, ongoing, the EDI support team brings up additional EDI partners, resolves issues as they arise, and maintains the currency of the EDI system. For questions and problems involving data content, functional business users on the support team act as the main trading partner contacts. On the other hand, for questions and problems involving syntax, IT support people are the main contacts. Various technical people on this team must be knowledgeable on either the EDI standard, the EDI translator, or communications. To limit the number of different people required on the support team, a combination of two or more are preferable. Planning and implementation tasks are discussed in greater detail in the next chapter.

7 Planning and Implementation

In this chapter we will discuss how to best plan for and implement EDI in your company by evaluating the potential benefits and savings in the various functional business areas, prioritizing and selecting the best EDI opportunities, and bringing the project to fruition.

Objectives

- Describe EDI as an intercompany project.
- Identify eight steps used to evaluate EDI opportunities within a functional business area.
- Discuss various perceived barriers to entry and appropriate responses to each.
- Describe the tasks and deliverables of the implementation plan.

EDI, the Intercompany Project

Implementing EDI, an intercompany system, takes a good deal more time and coordination than other internal projects. When developing the typical intracompany (within the company) system, it is fairly obvious who the major players are. The functional business manager usually takes the lead role in defining business specifications for the new system. These are passed to the Information Services (IS) manager who evaluates the required work and assigns system design and development tasks to one or more system analysts and computer programmers. When more than one user or business area needs access to the same information, the MIS group provides a link to the system's data files. Once the system is programmed and tested, it is considered live and usable by its various business users. The implementation of EDI differs in a few important ways.

First, the system information requirements are based on a standard with which the data-processing staff is often unfamiliar. This issue is dealt with by training technical people on the nuances of the EDI standard or buying an EDI translator to act as the standard interface. Nevertheless, there is still some programming to be done to develop a system link to the existing business applications.

Second, the system must contain control features over and above those included in internal computer applications such as:

- It must capture the information needed to develop and maintain a complete audit log of EDI activity.
- For outgoing transmissions, it must be able to generate a complete set of EDI standard control envelopes, comprised of interchange, functional group, and transaction set headers and trailers as well as complete and correct EDI transaction sets. For incoming EDI files, it must read and validate these same control envelopes in addition to verifying that the incoming transaction sets adhere to standard architecture and syntax rules.
- It should be able to reconcile counts it has retained on EDI activity with incoming EDI VAN reports and trading partner functional acknowledgments and produce exception reports of discrepancies.

Third, internal system requirements are typically based solely on intracompany considerations, while EDI requirements are more intercompany related. Data requirements come from both the standard and communications requirements from a VAN or a trading partner. In addition, business procedures are designed to satisfy partnership requirements. Because of the intercompany nature of EDI, staff members from different companies must come to agreements and coordinate activities and testing schedules in order to implement the project. As a rule, this introduces a great deal of complexity and requires a great deal of coordination. To facilitate this process, participating companies must carefully keep track of who is responsible for which internal tasks and who the counterparts are at the other companies. In addition, they often assign responsibility for coordination of the project to a staff member and require sign-offs on all tasks.

Evaluating the Business Area

As the concept of EDI most often enters the company either as a request from a trading partner (usually a customer) or as a solution to an internal inefficiency, it is most often focused on particular business transactions and in a particular business area. For example, the business manager may need to handle and process purchase orders more efficiently in response to a customer's need to receive product more quickly, as is the case with JIT manufacturing schedules, or he or she may need to speed up invoice approval by reducing the staff and time required to reconcile incoming invoices against open purchase orders and receiving information. In order to evaluate the costs and potential benefits of EDI, the best place to focus is on the functional business area. The deliverable of this evaluation process is called a *feasibility study.* The major players in this work are, first, functional business managers and users who analyze current business practices and determine the business person's processing and information needs for the EDI environment. Second are the MIS manager and data-processing support staff who evaluate the work and time required to implement the system changes needed to support the business processing and information access requirements. While performing the business area analysis, one must keep in mind that implementation of EDI itself results in direct benefits such as the elimination of mail time and key entry; all other benefits are indirect and are dependent on the development or enhancement of business systems and the changing of business procedures. Because of this fact, the first step is to discover exactly what is happening today by analyzing current systems and procedures in the functional area.

The following steps, illustrated in Fig. 7-1 should be performed for each business area under consideration for EDI implementation. Figure 7-2 shows the deliverables of each step of a business area feasibility study.

1. *Develop a list of the business transactions that are currently handled in the business area.* Business transactions here refer not only to paper business documents and forms but more broadly to any vehicle for sharing of the information needed to facilitate business such as telephone calls, telexes, facsimiles, paper reports, catalogs, lists, and oral transactions that act as backup to paper business documents. If you include only the traditional paper documents but inadvertently leave out all the supporting paper and oral transactions, you will not be able to create a true picture of what is happening in the business area today. Be sure to note transactions that circulate within the business area such as product requisitions which may be initially handled by a buyer and then authorized by a manager. Also, be sure to account for those that are used in other areas as well. For example, one copy of a document is sent to a trading

1. Develop a list of business transactions used in the business area.
2. Develop a list of the manual tasks performed in the business area.
3. Develop an information-flow diagram of the business procedures and a data-flow diagram of the systems in the business area.
4. Combine and revise the flow diagrams.
5. Develop a list of potential EDI trading partners.
6. Compute the cost savings you expect to realize from implementing the revisions of step 4.
7. Jointly determine the scope of the EDI project.
8. Jointly develop a business area recommendation.

Figure 7-1. Feasibility study steps.

1. List of the business transactions used in the business area
2. List of the manual tasks performed in the business area
3. Information-flow diagram of the business procedures and data-flow diagram of the systems in the business area
4. Combined flow diagrams with revisions to streamline business procedures and eliminate manual tasks
5. List of potential EDI trading partners
6. Computed cost savings after revisions of step 4
7. Jointly determined scope of the EDI project
8. Jointly developed business area recommendation for implementing EDI and phasing in EDI trading partners

Figure 7-2. Feasibility study deliverables.

partner, a second is filed in-house, and a third is forwarded to another department. Finally, include those that flow intercompany between yourself and your various trading partners.

2. *Develop a list of the manual tasks currently performed in the functional area.* This is best done by the people actually performing the tasks. Be sure to include not only the regular tasks that comprise the major part of a job but also those that are performed irregularly, such as exception-processing tasks, and infrequently, such as quarterly or annual tasks. Also, be

sure to consider all stages of the paper business documents as discussed in Chap. 3 on developing a cost justification. A sample list of tasks for incoming and outgoing documents is shown in App. A.

3. Step 3 has two parts. The first is to *develop an information-flow diagram of the current functional business area.* For this step, use both the transaction and manual task lists developed above to determine what is currently transpiring in the functional business area. Using the data-flow symbols shown in App. B, you can develop a picture of the current flow of information into, through, and out of the functional area.

Make note of the number of staff members and the time they need to perform the manual tasks identified above. If you can get figures on the error rates you are currently experiencing, they will help you to determine your current exception-processing costs. Include such errors as incorrect interpretation and key entry of business information, program malfunctions or inappropriate actions caused by entry of incorrect information, incorrect or late shipments sent or received, and incorrectly shipped and returned goods.

For the second part of step 3, *develop a data-flow diagram of the systems used in the functional area being analyzed.* You may already have a data flow showing all the functionality of the system as well as all data files which it either has access to or updates with new or more current data.

It is the combination of business information flow and system data flow that actually shows what is happening in the functional area. When combining the two, the technique is to replace the circle on the business flow that represents the computer system with the entire technical data flow that illustrates all the system tasks. We say that this "explodes" that one business flow circle.

4. In step 4 we examine the combined information and data-flow diagrams and identify opportunities to eliminate duplication of effort and to streamline and automate currently manual procedures. We begin by *developing a revised data flow for the business area* showing any residual manual tasks along with system tasks as they would exist with EDI and after re-engineering of current procedures. Be sure to include the following:

- Automated reading and interpreting of incoming EDI data if you are the receiver of the transaction, and automated generation of an EDI data file to replace the paper business documents if you are the sender
- Application link and translation software to map from the EDI standard to the file required by the application, or from the application format to the EDI standard
- Control mechanisms designed to track EDI activity by reconciling it to functional acknowledgments, and VAN control reports

Productivity tools such as on-line systems provide access to needed business information and have ad hoc computing and reporting capabilities. This may very well be the most important component of the revised flow as it shows how access to information will improve in the EDI environment. The tools provided here are often what sells EDI to the business community.

The revised flow shows the long-term picture, which does not preclude you from phasing in implementation of the various changes over a period of time. Without the big picture, you might very well limit future results by missing opportunities to integrate your internal systems or streamline procedures that are peripherally related to the area under consideration. Actually, the best way to develop the right long-term picture for your organization is to perform this same analysis and evaluation for several business areas. Then combine and revise the overall flow of business information.

There are two major benefits that you will realize. One, you will discover opportunities to streamline and eliminate duplication of effort between functional areas in addition to within one area. Two, you will discover instances where the monetary and resource investments needed to implement EDI in one functional area will be offset by major savings in another area. For example, customers who send orders electronically actually attain most of their savings by automating reconciliation of incoming invoices. Unless the organization were looking at the big picture, would the purchasing department be interested in spending the time and money to convert purchase orders to electronic form so that the accounts payable department could streamline procedures and "right size?"

5. *Develop a list of potential trading partners* with whom you want to initially implement your chosen EDI application. Also develop a list of partners for the longer-term rollout of the EDI program. You may want to use this list to survey those partners with whom you are interested in trading electronically to ascertain their experience or interest in EDI. This will help you to determine how fast you can reasonably expect to roll out your program and the savings you can reasonably expect to realize.

6. *Compute expected cost savings and required investments* as described in Chap. 3. From the results of this computation, you can easily determine the expected break-even point, the anticipated stream of ongoing savings, and whether and when to proceed.

Electronic data interchange can rarely be cost-justified in the short run, as setup and implementation costs are heavily front-loaded. Using anticipated volumes of EDI activity over time, you can compute expected savings for a period of years. Remember EDI savings happen in direct proportion to the amount of business conducted electronically. On the other hand, strong market factors may outweigh all other cost-benefit considerations. For example, pressure from a major customer often precludes the need for a vendor to perform a formal cost justification.

7. *Jointly determine the scope of the EDI project.* Using the requirements and expected savings on the business side, along with the cost of the technical effort needed to respond to these business needs, the scope or size of the EDI project can be determined. In general, the more automation and savings projected, the more chance there is that extensive and expensive enhancements will be approved by upper management. The most important thing is that the needs for both business and technical groups must be addressed in the project plan. Often, companies phase in their EDI implementations to more closely link expenses to expected results. In determining how to phase in new EDI transactions and new EDI partners, companies often survey their trading partners to ascertain experience or interest in implementing EDI. In this way, they can more accurately link internal support costs with potential savings.

Often, the overriding consideration in the initial implementation of EDI is whether it is required by a major trading partner as a condition for doing business or whether the company is playing competitive catchup. However, even when the implementation of EDI is a foregone conclusion, conducting the analysis outlined in the foregoing six steps and this step provides the business case to proactively build on the initial required implementation and to provide better and faster access for the business community to incoming EDI data.

8. *Develop a business area recommendation* as to whether and how to proceed toward implementation and then present the project recommendation to management to gain approval and funding to proceed. Focus on business considerations such as:

- Is EDI viable at this time?
- Are trading partners interested and ready to implement EDI with us?
- What procedural changes will be needed for the new environment?
- How will staff information requirements change?
- What new functionality and what data requirements will systems have to support in order to transact business electronically?
- How much funding will we need and over what time period?
- What savings can we reasonably expect and over what time period?
- What are the next steps?

Not discussed thus far, but very important in the planning stage, are the corporate and employee attitudes toward change in general and toward the introduction of technologies and automation in particular.

As we will see, staff attitude and degree of acceptance of change may be a major barrier to the entry of EDI into an organization. These can easily stonewall the EDI program unless confronted and responded to prior to implementation.

Objections to Entry into EDI from within the Company

Between August 1985 and July 1986, the First National Bank of Chicago conducted an EDI management survey. It asked questions about the respondent company; its current EDI activity, if any; and the perceived barriers to entry into the EDI arena. More than 21,000 surveys were mailed, eliciting over 1200 responses, or a return rate of about 6 percent.

The firms that responded ran in size from under $30 million revenue to over $1 billion. Of those, almost 60 percent currently had no EDI trading partners and more than 20 percent had fewer than five. The responses showed that the perceived barriers to entry into EDI were, in order:

- System cost
- Security
- Lack of standards
- Float loss
- Lack of training
- Company attitude

Today, more companies are doing EDI. Nevertheless, the list is much abbreviated. Currently, the top two are company attitude and lack of training. For those readers just now contemplating EDI, let's briefly respond to the first four to see why they are no longer of major concern. Then we'll address the last two to see how they can be eliminated as well.

System Cost

The perception was that there were extensive up-front costs to develop or buy the software and systems required to implement EDI.

While it is true that the bulk of costs related to implementing EDI is heavily front-loaded, these costs are not of main concern and are typically less than for other computer applications that we implement. Electronic data interchange system costs are directly related to the computer platform on which it resides. A small company or one that is looking at a fairly limited implementation can start out on a personal computer with a modest outlay of funds. Even for such a company, possibly needing a computer as well as translation software, a communications board, and protocol software, the cost of entry is not prohibitive. The total package would run between $3000 and $9000: from $1000 to $2500 for a microcomputer, from under $1000 to about $4500 for microbased translation software, and from $1000 to $2000 for a communications board and communications protocol software. At least one vendor is even offering a leasing package comprised of computer, software, and printer for one low monthly fee. This may be an interesting solution for the small company looking for a turn-key EDI system. Of course, if the company already has computer hardware, the cost would be considerably less. A larger company, or one that is planning an extensive EDI implementation, may still opt for a micro front-end as a start-up solution and plan to move to a larger platform when its EDI activity justifies it. This keeps costs in line with payoff. The cost of a midrange EDI translator is somewhere between $11,000 and $25,000, which can be cost-justified because it processes considerably faster and is much more feature-rich than its microcomputer counterparts. Mainframe computer translators are more expensive yet, running from $35,000 to greater than $50,000. As you move higher into the range, the price includes additional modules that work in conjunction with the translator to provide additional functionality on which you would otherwise be devoting development funds. For example, you may purchase a module that allows you to automate your EDI controls by using the basic control information collected by the translator to identify exceptions and reporting them to the appropriate parties. Or, you may purchase a module that allows you to handle EDI in a real-time manner with an on-line component.

While these up-front costs are certainly not inconsequential, they proved not to be prohibitive. For a company that was also reengineering business procedures and systems, EDI was often the smaller component of the cost.

Security

The perception was that there was a marked lack of data security with EDI since access is readily available to all and data can potentially be read or changed during transmission.

Back in the mid-1980s, business people were less comfortable than they are today with conversion of paper business documents to a file form. They pointed out that these files were more easily accessible to improper viewing than paper documents locked away in file cabinets. While this was potentially true, security procedures could be put into place to allow only authorized personnel to inquire or modify data. In fact, various levels of security and data access can be built into the system so that each log-on ID and password is limited to only the level of access required by that person.

Business people were also uncomfortable with the idea that data can be read and possibly changed while in transit between partners. In cases where this is truly an issue, encryption (scrambling of data) can assure that data cannot be understood by the casual observer during transmission. In addition, authentication can assure that any changes made during transmission would be able to be detected by the receiver. While both of these techniques are available for use, both require additional software and hardware and a good deal of management and coordination to maintain encryption or authentication keys. In addition, new procedures and system comparisons must be performed. Consequently, both are used only when it is determined that they are required to guarantee the needed level of security.

Today, there are very few instances of companies using either encryption or authentication. Nevertheless EDI data are considered more secure than the typical paper document because EDI business transactions can be backed up regularly and accessed randomly.

Another security misperception was that if trading partners communicated directly with you and retrieved data from your computer, they posed a security hazard to the integrity and secrecy of your internal files. In fact, this fear was not based on reality, as EDI is a *send-only* facility. As illustrated in Fig. 7-3, this is the way it works. One company calls the other's data center, identifies itself, and says it is ready to receive. The other company, satisfied that the first has identified itself sufficiently well to its system, *sends* the data to which the receiver is entitled. Actually, there is no direct entry into the computer and no pulling of data at all. Sometimes, an EDI VAN is perceived as a safety buffer between a receiver and a sender's computer. While this is not a necessary buffer as explained above, using a VAN often makes the trading partners feel that their data are more secure.

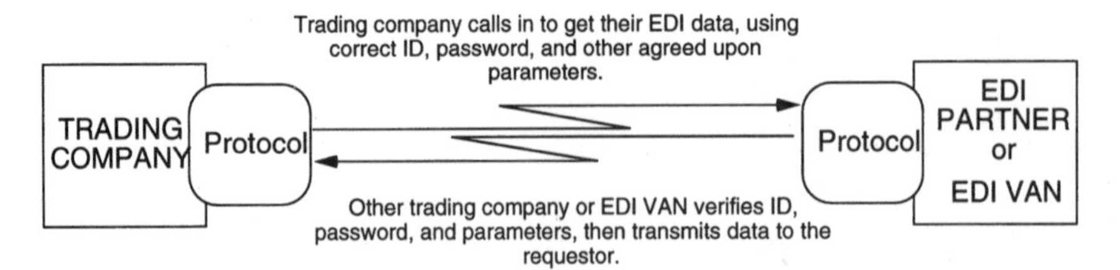

Figure 7-3. EDI as a send-only facility.

Standards

The perception was that the standard was in too much flux and had not gathered broad enough industry backing to warrant the implementation effort.

Today, the ANSI X12 standard, supporting cross-industry transaction needs, has gained broad support from a large number of industries. Often an industry new to EDI finds deficiencies in the standard as it currently exists. They typically clear up their concerns through maintenance requests to provide the changes they require. The basic structure and data content have held up remarkably well because of the natural homogeneity of business information needs between companies, industries, and nations. For companies planning on implementing EDI today, the X12 standard is very stable and information-rich.

Float Loss

The perception was that float, the time period between when a check is written and when funds are collected, will be eliminated if EDI, in the form of electronic funds transfer, is implemented.

I believe this issue only made the top six concerns because the original study was conducted by a bank and was sent to a large number of financial business people who may have been concerned more with float loss than with EDI. While it is true that float is eliminated if payments are made electronically, it is not true that vendors have access to funds any sooner than they would have in the paper world.

Let's take the example of a payment that must be made within 10 days of the invoice date in order to take discount terms. In the traditional environment, the check would be mailed on day 10. The buyer would experience a 3-day float period during which the check was in the mail and an additional 2- to 3-day period until the check cleared. Accordingly, the buyer would not even have to maintain balances to cover the check until day 15.

In the EDI environment, balances may be transferred electronically from buyer's to seller's accounts as soon as authorization to pay is received by the bank. While there is no float period in this transaction, the buyer can authorize payment on or as of a specified date. This allows the actual payment to clear on the same day as previously. Some companies renegotiate their payment terms when they implement electronic funds transfer (EFT). The vendor may choose to extend the discount period an additional few days to make up for the 2- to 3-day clearing period. The bottom line is that experiencing the loss of float does not necessarily mean experiencing a loss in cash flow.

Training

The perception was that no one in-house knew enough about EDI or data standards to oversee the implementation effort.

It was true that most companies had few, if any, EDI-knowledgeable people on their staffs. In light of the fact that few robust EDI translator products existed, lack of expertise in the standards was a true obstacle to implementation. Even though there were educational opportunities, there were few compared to those offered today by EDI educational organizations, schools, standards groups, industry groups, and various business focus groups. As a matter of fact, there are a multitude of educational opportunities today. Starting with product education, almost all translation software vendors offer classes on their products' features and how to most effectively use them. On more general topics, there are many conferences each year that are either entirely or partly devoted to EDI. Many invite software vendors and third-party providers to display their products and services. Most have educational workshops on various EDI and electronic commerce topics. In addition, there are many experts in the field who provide on-site consultation and education. Finally, most companies have at least one trading partner that has already implemented EDI.

If all these educational opportunities exist, how is it possible that lack of training is one of the major barriers to implementation today? The reason is that when lack of education is cited, it refers not to technical staff but to the business community. It has always been intuitively obvious that EDI education is valuable for technical people. Not only are IT managers willing to educate their people, but the people themselves are requesting the education. Attendance is usually assured at classes that cover the components of an EDI system, the EDI standard, and various technical aspects of implementing EDI. That is why most classes offered in EDI are slanted toward the technical side. Even when the information technology staff can install and run an EDI translator and communications software, the business people still rarely see their part in the project or understand the tremendous impact EDI and other forms of electronic commerce can have on the organization. What's more, this is true not only for low-level business users but for departmental managers, vice presidents, and presidents and CEOs as well.

There are two obstacles standing in the way of educating the various levels of business people. One, most business people still see EDI as a technical capability, not a business initiative. Therefore, they do not even know that they should be interested in education. Two, even when a business person attends a class, he or she is almost always the business user. Unfortunately, recommendations to analyze the current business environment, reengineer procedures, and institute change are not areas in which a business user can exercise any control. Furthermore, since influence in an organization flows uphill a lot less quickly and effectively than downhill, these newly initiated business people have difficulty in impacting their company's preexisting attitudes and directions. The next barrier, company attitude, is very closely related to lack of training.

Company Attitude

Upper management believes that EDI is a technology. As such, it is reasonable to expect that it be implemented solely by information technologists and that it will impact only the system environment of the organization.

We all know that without the support of upper management, no new initiative can happen. First, it will get no funding; second, staff will not be allowed to devote time to it. Here the issue is not that top management does not support EDI, it is that the part of the EDI project that they know about and support, the technical part, is merely the tip of the iceberg. The business changes needed to take advantage of electronic commerce and position the company for real savings cannot happen because executives are unaware that EDI is a business initiative that will impact the entire organization.

Over time we see that successful EDI implementations occur when there is not only awareness but top-down pressure to succeed with quantified expectations of results. For example, "We intend to have *X* partners up within the next year, which represents *X* percent of business." When this type of top-down pressure exists, it is often the result of much work done by middle management to build the case for EDI through the process we described for the feasibility study and through research into what others are already doing. The result is to prove to upper management that savings can be realized in-house.

The question is, "How is it that this is accomplished in some companies but doesn't ever occur in others?" I believe that the answer is, "In those companies where EDI is positioned as a relevant and valuable tool to assist the business community to reach its corporate, departmental, and job-related objectives, EDI is strongly supported by all levels of business. However, where EDI is positioned as a new technology, no one outside of the IT department supports it."

Since EDI is generally introduced into a company at the middle-management level, middle managers are charged with passing the new requirement on to upper management. And, because they are not familiar themselves with the potential benefits of EDI, they typically present it as the introduction of a new technology and stress only hardware and software needs. The relevance of the new technology to attain corporate objectives often eludes middle management and causes the wrong message to be passed on to top management. This is turn leads to EDI implementations that are technologically appropriate but are not tied to any real business needs. Here is a suggestion on how to eliminate the EDI program that is implemented in a vacuum. It is based on the premise that *people are always more interested in things that directly relate to them.*

For top executives, this translates into, *"What will help me to accomplish my corporate objectives?"* If our corporate goal is to gain market share, what will help me to take business away from competitors and open new market segments? If our goal is to increase profit margin by lowering internal costs, what can I do toward those ends?

For middle-level managers, the questions are, *"How can I assure accomplishment of our departmental objectives and maintain good relationships with trading partners?"* If we are feeling trading partner pressure because innovative competitors can offer a higher service level than we can, what can we do to counter that? If we have difficulty in providing timely and accurate responses to trading partners with our limited resources, how can we use my staff more effectively and efficiently?

For business users, the questions are, *"How can I accomplish my job goals in the most efficient manner?"* If I have quotas to meet, whether internally oriented, such as answering a predefined number of customer service calls or processing a certain number of business documents, or externally oriented, such as selling a predefined dollar amount of product, "How can I assure that I am able to meet the expectations in the allotted period of time?"

When EDI is presented to the various business people as an answer to their business needs, they not only buy into its introduction but become strong proponents as well. Sometimes bringing in outside expertise provides the credibility needed to get the attention of the various business people. By focusing on the business aspects of EDI and tying the implementation of EDI to your company's reengineering plans and corporate, departmental, and individual goals, you will find that business attitudes will change over time.

In regard to attitudes, there is one other important point to be made. Those organizations that have as corporate goals the attainment of high productivity and internal efficiency are more likely to succeed at EDI than those who have not yet adopted the "lean and mean" philosophy. This is because department size tends to shrink in those areas where EDI is implemented. If that is not an end to be desired, managers of large departments, typically enjoying high status based on department size, tend not to become proponents of the change. In general, unless high productivity is the ultimate goal, EDI will have a tough time gaining a foothold in just those departments in which it is needed the most.

Because the objections listed above commonly act as barriers to implementation or serve to keep EDI focused narrowly on only one application, you may find the responses given here to be interesting and useful to you within your organization.

EDI Implementation Phase

The implementation phase follows the planning or feasibility stage. Figure 7-4 shows the additional deliverables needed to develop a corporate implementation plan. In order to generate these deliverables, the following tasks need to be accomplished:

- Prioritizing business area feasibility studies based on readiness and potential savings
- Agreeing on information requirements, business procedures, and EDI technical arrangements with trading partners
- Evaluating and selecting translation software products and communication options
- Defining application link processing requirements and specifications for its development
- Selecting pilot partners
- Developing and running the pilot program
- Defining a rollout plan for the project and establishing program goals

As discussed in Chap. 6, in order to accomplish these tasks, it is necessary to put together an implementation team. This team often contains representatives from the same business areas as were on the steering committee but usually of a lower level. For example, while the steering committee may have a vice president from the sales and marketing area, the implementation team will have a sales manager. In addition to the functional business people, this team will also have representatives from other departments such as:

- Accounting and legal representatives, as needed
- Information technology support staff
- Communication support staff
- Human resources representative

1. Prioritized functional business areas by computed payoff
2. Information requirements or availability for business transactions that will be converted to EDI
3. New business procedures to support the EDI environment
4. Selection of EDI translation software
5. Decision on communications option, direct or through an EDI VAN
6. Selection of an EDI VAN
7. Definition of application link requirements
8. Selection of one to five pilot partners
9. Definition of the EDI pilot plan steps and milestones
10. Assigned responsibilities for pilot program
11. Establishment of EDI goals and rollout plans

Figure 7-4. Additional implementation-stage deliverables.

Sometimes, functional managers and data-processing and communication counterparts from trading partner organizations are asked to participate as well.

Before examining the various tasks that must be performed by the implementation team, it may be helpful to review the deliverables of the feasibility study (see Fig. 7-2) and the list of additional deliverables of the implementation plan (see Fig. 7-4). Implementation tasks can be grouped into four categories: trading-partner–related; data-processing application, application link, and EDI translator support; communication support; and full committee. Let's examine each category in some detail.

Trading-Partner–Related Tasks

Up until this point most of the EDI-related tasks are internally focused. There is little interaction between company staff members and trading partner organizations, except possibly when compiling the list of potential EDI trading partners for the business area feasibility study. Now, however, it becomes necessary to build on the already good working relationship with trading partners with whom you wish to do EDI. Working closely with trading partners' functional business managers and IT counterparts will ensure that pertinent business and technical issues are discussed and resolved prior to the implementation of EDI. Following are some of the important tasks in this category.

Finalizing Business Specifications for Internal Business Systems. As a result of the feasibility study, your implementation team's functional business managers have recommended converting certain transactions to EDI and have identified the basic information available from or needed by the existing internal business systems when handling those transactions. Prospective EDI trading partners may very well provide input on the priorities you have assigned to the EDI transactions. For example, if you are a vendor preparing to receive EDI purchase orders and return invoices, your customer may prefer instead to send purchase orders but receive advance shipment notifications. Trading partners may also have input on the information you anticipate sending and receiving via EDI and on how it is represented in the standard. By comparing your information requirements against those of your partner, exceptions easily surface and both partners see where their business systems or application links need to be upgraded. Any changes made during this exercise should be incorporated into the business specifications and eventually into programming specifications for the business system or application link.

Let us digress just for a minute to discuss the terms information and data. *Information* is used here to refer to what the business person needs to perform his or her tasks. Information is comprised of words and sentences and is formatted in an easy-to-read print format. *Data* is used to refer to words, numbers, and codes used in computer software processing and the EDI standard format. In general, business people are interested in information, while IT people are interested in data. When dealing with trading partners, business counterparts from the two organizations discuss information needs. Technical counterparts discuss how information needs can best and most efficiently be represented in the EDI standard. For example, a business person needs ship-to name, address, and contact information in an incoming purchase order. The technical staff may suggest receiving only customer ID in the standard and accessing the additional name and address information from its own customer files. The final result is minimization of the amount of data that are transmitted while giving the business people the more narrative and easy-to-read information that they need.

Discussing information and data needs at this early stage points out potential problems, such as the absence of data fields that the receiving trading partner may consider mandatory. Based on the information and data agreements reached between the partners, either one may need to enhance its business application and application link. It may need to provide or accept additional fields and may even need to add processing steps for new information requirements. Any changes made during this exercise must be incorporated into both the business and technical specifications.

Reaffirming Conditions and Terms for the EDI Environment. While paper transactions carry with them a reaffirmation of terms and conditions in small print on their reverse side, no such vehicle is available in the EDI data stream. Consequently, many firms reaffirm terms and conditions in a separate contract or in an EDI rider to an existing contract. Either way, both parties sign and date the agreement, and it remains in effect until either party wishes to change it. Often, this rider is merely a reiteration of the existing terms and conditions with only some timing expectations changed for the electronic environment. For example, instead of the traditional 10-day period during which a vendor must respond to an incoming purchase order, a 24- or 48-hour period is substituted.

Developing the Pilot Program. One of the main differences between EDI and other internal projects occurs during the testing phase. With most projects, business users specify their needs, technical people develop programming specs and write the system, and then together they test to determine if the system fulfills the business needs and sign off on the accepted system. With EDI, this same thing happens during the initial implementation. However, this is only a first step. After initial acceptance, the EDI project must undergo a testing and accepting phase with each new partner. Most companies that anticipate bringing up many EDI partners opt to standardize their implementation approach by designing a pilot plan that they use during initial testing with each partner. The contents of the pilot plan should be discussed with each prospective EDI partner. The pilot plan itself contains all the tasks needed to set up, test, and then evaluate the hardware, communications, software, and procedures with each pilot trading partner. All tasks are performed with each new partner. Each plan task is a specific activity that must be performed, evaluated for success, and signed and dated by the responsible party to show its successful completion. In a structured plan, sometimes various related tasks are grouped into milestones. It is best when each task and milestone is defined in terms of the criteria for its success as this makes it easy to determine when and if it has been successfully completed. For example, "complete communications testing" would be too vague as a pilot task. However, it may very well be a pilot milestone containing all the communication-related tasks needed to assure both partners that communications are working properly; that all characters can be received and understood; that the EDI standard file can be generated, transmitted, and received intact; and that the standard file contains all the agreed-upon data fields. This milestone may contain the following tasks:

- Each partner must make contact with the other via telephone connection, using a specified communication protocol and line speed, and receive a "ready" through the handshake of protocols.
- Each partner must make contact as before and transmit a predetermined data stream containing all letters, numbers, other characters, and spaces.
- The receiving partner must analyze the incoming data stream for completeness and correctness and report the results to the other.
- Each partner must make contact as before and send a predetermined EDI standard data stream containing data in all the mandatory and optional fields preagreed to by the two partners.
- The receiving partner must analyze the standard data stream for adherence to standard syntax rules and report the results to the sender.

Other examples of milestones are

- Interpretation and processing of transaction *X* (e.g., purchase order) into receiving computer application (e.g., order entry).
- Parallel-test paper and electronic versions of transaction *X*

To facilitate administration of the plan, develop a form that contains the complete list of milestones and task names in a column going down the left side of the page and associate the following additional information with each task:

- Task or milestone number
- Task or milestone description
- Anticipated completion date
- Actual completion date
- Party responsible for task
- Comments

Be sure that you assign responsibility for administering the pilot and tracking completion of pilot tasks to a specific person. As each task is successfully completed, its status and date should be recorded and signed off by the responsible party. The plan administrator should coordinate that effort. Once all the tasks and milestones have been successfully completed, the pilot partner can be converted into production mode.

If problems are encountered during the pilot phase, the narrative in the comments fields should shed light on what they were and the reason they were encountered. By discussing status with the parties responsible for completion of the tasks, you can get the pilot moving again toward successful completion.

Data-Processing Support Tasks

Business specifications, describing additional functionality needed by business systems and accounting requirements, describing the collection of adequate control information and development of an EDI audit trail, are used by the data-processing staff to define enhancements to existing systems and to help design new systems. Data-processing application specialists, intimately familiar with existing computer applications, evaluate these new specifications and requirements in terms of existing application functionality. Based on their evaluation, they are asked to prepare an estimate of the time and effort needed to incorporate changes into existing and new computer programs.

These time and effort estimates become part of a recommendation that also contains alternative translation and communications options. The recommendation is presented to the implementation team who evaluates it and determines the best way to proceed. If internal development is needed, the application specialists will be charged with developing programming specifications, then programming and testing the required modules. In addition, if the decision is to use an EDI VAN, they will be asked to prepare a list of data-processing–related VAN requirements and questions. This list will become part of the request for proposal (RFP) to be sent to prospective EDI VANs. Responses to the requirements and questions in the request will help the implementation team to evaluate VAN candidates.

Communications Support Tasks

Communications specialists have several tasks for which they are responsible. The results of these tasks provide the basis for the communications solution they will recommend to the implementation team.

Initially, this group is called upon to evaluate the company's intracompany and intercompany communications capabilities. While the focus in EDI is on the intercompany aspects of the interchange, once data have been received, they often have to be disseminated to various locations through an intracompany network. Externally, each trading partner needs to support telecommunications to the outside. The typical EDI implementation uses a public, dial-up telephone line or permanent leased line with an acceptable error-detecting communication protocol and a line speed sufficiently fast for the anticipated volumes of data. Internally, the company needs the ability to move data from one data center to another and provide access to data between one business system and another within a data center.

Using these evaluations, communications specialists prepare a time and cost estimate for purchase, installation, and testing of additional hardware and software. In addition, using projected data volumes and number of trading partner transmissions, they evaluate the adequacy of the following:

- Computer storage and processing capacity
- Number of telecommunications ports to the outside
- Number of staff in the department
- Operational hours per day and days per year to determine the impact EDI will have on the operations department

Many companies find that the additional workload and responsibilities of handling and controlling multiple EDI transmissions from and to multiple trading partners would be detrimental to the working efficiency of their operations departments. They, therefore, recommend employing an EDI VAN as a communications intermediary. They base the recommendation they make to the implementation team on a list of the pros and cons of direct VAN communications. Over 70 percent of all EDI data travels through EDI VANS today because of just this type of evaluation.

If the recommendation suggests use of a VAN, communication specialists should include in their recommendation a list of communications-related requirements and questions. This list will be included in the request for proposal to be sent to prospective EDI VANs. Responses to these requirements and questions will also help the implementation team to evaluate VAN candidates.

Full Committee Tasks

While individual groups on the team make recommendations on different aspects of the EDI implementation project, the full implementation team holds meetings at which they evaluate those recommendations and make final decisions.

Specifically, recommendations presented by the applications support group and the communications specialists are discussed in terms of costs, time and effort, and alternative solutions. If the use of an EDI VAN has been recommended, a request for proposal document is generated by the group and sent to prospective VAN candidates. Included in this document are pertinent business, hardware, and software facts as well as the questions developed by the two technical groups. Also included are questions on the VAN itself such as:

- What is the VAN company's EDI experience and who are their existing customers?
- What are the communications protocols and line speeds that it supports?
- What are the hours and geographical range of its services?
- Which EDI and electronic commerce services does if offer?
- What other EDI-enabling software products does it offer?

Responses to the requests are received by return mail and are evaluated by the implementation group, who may request that one or several candidates visit and make presentations to them prior to making a final decision.

The proposed pilot program is also discussed at a full group meeting. In particular, this group determines if the proposed start date of the pilot is realistic in terms of successful completion of all the tasks leading up to it.

Individual members also examine the pilot plan to verify that sufficient time has been allotted for each task or milestone for which they are responsible. While the pilot is usually administered by one member of the implementation team, it should be considered a group effort. Additionally, team members discuss and agree upon the success criteria for the pilot. Typically, success is defined in terms of the number of new EDI trading partners or amount of newly converted business in a specified amount of time. At the end of the pilot's duration, its success or failure is determined by evaluating the results in terms of the agreed upon criteria. Lack of success is often an indication of coordination problems as issues may arise in any one of several areas; communications, data processing, or business. The implementation team should monitor any issues that arise and make appropriate changes to the pilot tasks and their anticipated completion dates.

Summary

The EDI planning process cannot be overemphasized. Evaluating the uses and benefits of EDI in various functional areas provides an excellent first step toward incorporating EDI into the strategic plans of an organization. Each business area feasibility study uncovers additional opportunities to generate or receive timely and accurate information that can improve the overall profitability and productivity of the business. Bringing the right people together during the implementation process can mean the difference between a well-planned development project and constant loose ends and backtracking during the process. To develop full business and technical specifications and to complete the implementation tasks, functional business managers and users, application software specialists, communications specialists, and legal and auditing representatives should all be represented during the implementation process.

A predefined pilot plan takes the project from the implementation phase to production mode. In EDI, the initial pilot is used to bring up each new EDI partner, starting with three to five in the initial pilot program. Since the final goal is an EDI trading community comprised of all the key trading partners, it is particularly effective to monitor the initial pilot program to be sure that all the tasks needed for successful implementation are defined in the pilot plan and completed in a timely manner during the testing phase. Any inefficiencies should be eliminated to ensure that subsequent partners can be brought up quickly and with minimal effort.

If the EDI planning process has been handled properly, the marginal work to add each new EDI trading partner should and will decrease over time, which, in turn, will bring the company's EDI goal that much closer to fruition.

8
EDI in the Retail Industry

In this chapter we will focus on two sectors of the retail industry: grocery (food products) and general merchandise (mass merchandisers). In particular, we'll discuss each sector's economic environment, the impact it has had on its mode of conducting business, and the systems that it has developed. We have chosen grocery for two reasons. One, it started early enough to require that it develop its own industry-specific standard. Two, its EDI vision was broad, with a planned implementation of a variety of EDI transactions in several functional business areas. We'll discuss both the traditional grocery replenishment approach, where shelf space is controlled by the grocer who orders and gets periodic delivery from the manufacturer, and the direct store delivery approach, where shelf space is controlled by the manufacturer who replenishes the stock via a delivery truck route to the retailer's back door.

For general merchandise, we have selected mass merchandisers because of their size and influence in the marketplace and their need to cut costs and maintain high levels of customer service. As major customers, they exert tremendous pressure on their vendors. Because of competition, they need to offer products at the lowest prices and provide the highest level of service while still retaining profitability. All this leads to a higher growth of EDI in the mass merchandise segment of the retail industry than in any other segment or industry.

After discussing the background and basic EDI implementations of each segment, we'll focus on the programs they have each put into place to improve productivity and efficiency throughout the entire business cycle, from replenishment of product in warehouses and distribution centers to replenishment in stores. In the grocery industry, this program is called *efficient consumer response* (ECR). In the retail industry, this is called *quick response* (QR). In both instances the program requires the active participation of all players in the replenishment cycle, suppliers, distributors, brokers, and retailers. Likewise, both encompass sweeping changes in the way business is conducted and new technologies and systems to support the business. As part of the QR discussion, we'll focus on the apparel industry to highlight the innovative QR programs some manufacturers and retailers have implemented to maximize efficiency, decrease costs, and raise customer service levels.

Grocery

The grocery industry became interested, in the late 1970s, in identifying the cost and savings impact it would experience if it traded electronically with business partners. It hired the consulting firm of A. D. Little to evaluate the potential benefits of electronic trade for the industry.

The resulting study reported that the potential for benefits within the grocery industry (excluding the direct store delivery segment) was high. By converting only 50 percent of the normal paper transactions in the full procurement cycle to electronic form, direct savings of $84 million could be realized through reduced clerical costs and an additional savings of $128 million could be realized through enhanced systems and streamlined business procedures. These benefits would accrue to the supplier, the broker, and the customer in approximately 45, 13, and 42 percent, respectively. In the long-term, indirect savings of up to 3 times the size of these direct savings could be attained by developing additional computerized business applications, automating currently manual tasks, and integrating the new systems into the business environment in such a way as to make the most effective use of the timely and accurate data available through EDI. The procurement cycle transactions referred to in the study are purchase order, shipping advice, invoice, payment advice, and purchase order changes and adjustments. Additional transactions which would facilitate complete electronic processing of the purchase-deliver-pay process are price change announcements, price protection announcements, and promotions.

The study provided specific findings, recommendations, and predictions. It found that in order to link the manufacturer and distributor electronically, each would require an EDI system. Such a system would not require large capital costs to get started. While EDI at that time was technically feasible, the major impediment was the lack of an industry-endorsed message standard. Communication standards were also needed; however, the study found that they were not nearly so critical. The study called for the development of a message standard for the targeted transactions and for pilot implementations using the newly developed standard transactions to demonstrate that business could be transacted electronically with them.

From this beginning, the key companies of the industry moved ahead, customer-driven by such major retail chains as SuperValu, Giant, and Ralph's and with strong participation by major manufacturers and brokers.

These companies' first task was to develop the message standard. For this effort they enlisted the help of the Transportation Data Coordinating Committee (TDCC), which had developed the data standards for the transportation industry. The first version of the resulting Uniform Communication Standard (UCS) was published in April of 1982, along with the final report of the pilot study. The UCS was developed as a variable-length standard with syntax rules and control envelopes similar to those used in the transportation industry transactions. For communications, the bisynchronous 2780/3780 protocol at 2400-baud rate was the standard adopted because of its already widespread use among companies. In addition, an 18 hour/day *operational window* was specified during which each company was required to be active, i.e., listening for and able to receive incoming transmissions.

During the remainder of 1982, the initial pilot companies, using the TDCC as a resource to assist in contacting other companies, tried to attract additional trading partners to this new way of conducting business. At the beginning of 1984, the Uniform Code Council (UCC), formerly the Uniform Product Code Council, was given administrative control of the UCS standard in addition to the control it already had over assigning and maintaining the Universal Product Codes (UPC) that were used extensively to identify grocery products.

Very early on it was obvious that the standard would not be static but would periodically require maintenance. In order to support the ongoing changes to the standard, an advisory group set up a maintenance committee. This committee, named the Standard Maintenance Committee (SMC), was made up of volunteers from companies in all segments of the industry. It has kept tight control over the standard, evaluating all new requests on the basis of real business need. Because of its efforts, the standard has remained quite stable over time, changing only when enhancements were necessary to facilitate a business requirement. In practice, the standard is updated every 6 months. Each company using the standard is expected to incorporate the latest changes into its system within the next 6 months. In this way, everyone stays current in the standard and there is no need for any one company to support multiple past versions.

In order to become active in EDI, each grocery industry company was required to purchase, from the Uniform Communications Council, a communications ID which uniquely identifies them in their EDI transmissions. Communications IDs were priced commensurate with company size. The fees obtained through their purchases were used to support the future development of EDI transactions and to maintain the standard. More recently the use of other identifiers has been allowed. Companies have selected such unique numbers as their modem telephone number or Dun and Bradstreet (DUNS) number to serve as EDI identifier. These have become popular as EDI identifiers because they are unique and have no setup costs associated with them.

Most of the companies in the original grocery industry pilot decided to transmit directly with their trading partners. However, SuperValu opted immediately to use McDonnell Douglas as their third-party VAN. It was established from the beginning that any third party that wished to provide EDI services to the grocery industry would have to support an *open network*. Specifically, this meant that the VAN would support EDI transmissions on behalf of its customer, even when the other trading partner was not a customer. In order to do this, the VAN would be set up in such a way as to accept data from a noncustomer on behalf of a customer, requiring no special log-on sequence. And, it would be able to transmit data to a noncustomer on behalf of a customer via an autodial facility. These scenarios are illustrated in Fig. 8-1. Such a mandate existed in order to assure that every company wishing to do EDI would have the ability to do so without restrictions by trading partners or third-party agents. If EDI was to become the de facto standard of the grocery industry, it would have to work for everyone, regardless of size, computer platform, and system sophistication.

Many wholesalers and retailers installed their EDI systems on microcomputers, which were either stand-alone units or a front-ended midrange or mainframe computer. Many large manufacturers, distributors, and retailers who have business applications on midrange or mainframe computers prefer to have their EDI translators reside on the same platform. Those involved in the original pilot were forced to develop their own EDI translators, as no products existed at that time to provide the link between the business application and the UCS standard. Today, almost all EDI-active companies in the grocery industry are using a third-party VAN. Likewise, most have purchased EDI translator packages to replace the internally developed translators they originally installed.

Electronic data interchange activity today still centers around the purchase order and invoice; however, there are several other transac-

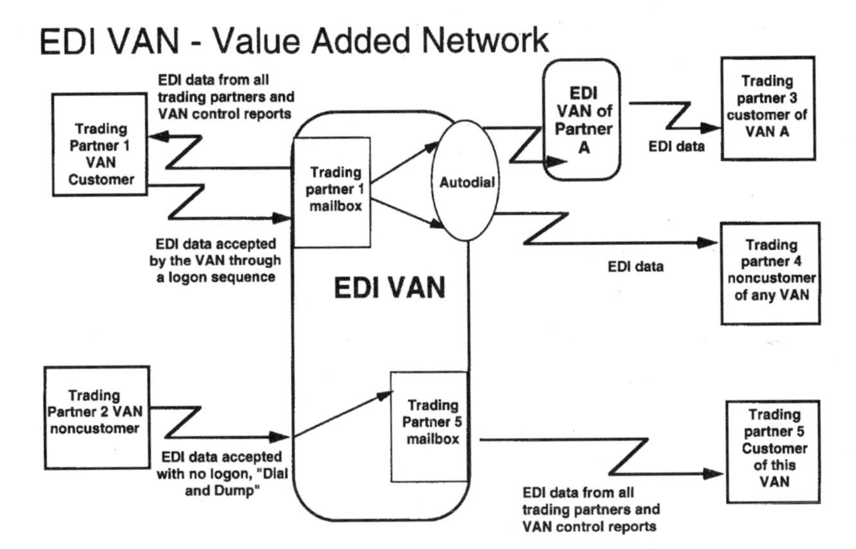

Figure 8-1. Open network for an EDI VAN.

tions beginning to be used more extensively. We'll cover those in the discussion on ECR. In addition, large manufacturers have developed an extensive electronic dialogue with the public warehouses in which they store product. The standard used for these transactions is called the Warehouse Industry Network Standard (WINS), which was developed in the same manner as UCS, utilizing the same data dictionary and the same syntax rules. Compatibility between the two standards allows information to be passed directly from a transaction set in one to a transaction set in the other.

The EDI data flow shown in Fig. 8-2 is possible with the targeted transactions of purchase order, shipping advice, invoice, payment advice, and purchase order changes and adjustments. Here, the purchase order is generated by the retailer using data from internal files and a computerized purchasing system. The data are passed to the EDI translator in a predefined, fixed-length file. The translator then moves each of the fields from the fixed-length file into the variable-length UCS standard, appends and prepends the appropriate control envelopes, and prepares to transmit the full file either directly to the manufacturer or through an EDI VAN on its way to one or many trading partners. The manufacturer receives the UCS-formatted data stream, verifies that it is correct and complete according to the syntax rules of the standard, and translates it into its own fixed-length file format through an EDI translator as well.

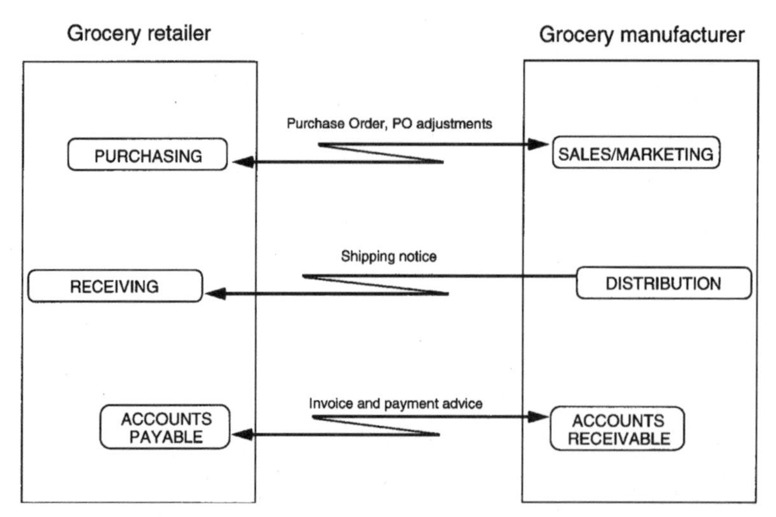

Figure 8-2. EDI flow between grocery manufacturer and grocery retailer.

The order is then processed by the order-entry application, which verifies available credit against internal customer files and product availability against inventory files. This application uses the purchase order information to generate an internal order, which it passes to the appropriate division, distribution center, or public warehouse for additional processing and eventual shipment.

Using most of the same purchase order information, along with corrections to reflect actual shipment quantities and delivery date, the manufacturer generates an advance shipping notice which it transmits to the retailer. The retailer uses the shipping notice to prepare for the coming shipment. The manufacturer can optionally generate a bill of lading for its carrier to notify them to pick up the shipment. The carrier, in turn, uses the bill of lading information in conjunction with standard freight rating tables to compute the freight bill. It generates an electronic freight invoice and returns it to the manufacturer.

In the event that the product is being stored in an outside warehouse, a copy of the EDI order is transmitted to the warehouse. This is used by the warehouse to stage the order for pickup by the carrier. In addition, various other EDI transactions travel regularly between the manufacturer and the warehouse to track the flow of inventory. For example, the manufacturer notifies its public warehouse when it plans on shipping additional product. In turn, the warehouse returns notification of receipt and condition of the product. Not only do these transactions help both parties to control their inventories more accurately and in a more timely manner, but they also permit the manufacturer to fill incoming orders from public warehouse stock days before it would have been available with the traditional paper and telephone environment.

Finally, the invoice is generated using the initial purchase order information along with actual shipping quantities. It is transmitted as soon as the order is shipped to the retailer. The retailer reconciles it against its open purchase order file and its product receipt information, and authorizes payment. Payment is made either by paper check or via EFT accompanied by payment remittance notification. In addition, changes and adjustments to purchase orders are transacted between trading partners as needed and processed automatically by the application.

What benefits can be realized from use of these electronic transactions? An aggressive retailer can realize both lead time and safety-stock reductions by shortening the order cycle. George Klima, former head of the SuperValu EDI program, predicted a savings of $600,000 per year for his chain. In general, there have been few reports of actual savings from manufacturers. In large part this is because most did not have a clear picture of their pre-EDI costs and therefore could not compute their post-EDI savings. However, there were other factors as well. For one, most companies implemented a less than complete set of EDI transactions with only a limited number of trading partners. Two, when EDI was in its early stages, many companies lacked the internal applications which could fully automate manual procedures. Today, the grocery industry is known for its extensive use of technology to improve productivity. However, there is still a stumbling block for maximum benefits from EDI. That is, while computerized applications exist in almost all departments, they are not always fully integrated with one another. The result is that there are gaps between automated systems where data are stopped, printed, and rekeyed into subsequent systems. Another result is the lack of continuity of the most current information throughout the entire organization. As the percentage of EDI business grows, benefits are undoubtedly higher. For manufacturers, this is happening because all the largest retail chains have converted to EDI and many of the medium and small chains are in the process or poised to go. For retailers this is happening as they continue to pressure vendors to support EDI. In fact, today, the focus has moved from just EDI to EDI plus other technologies such as bar-coding, shipping container marking (SCM), and scanning and to new business services that can streamline all facets of the business relationship. What's more, the focus has moved from optimization for each member of the supply chain to optimization throughout the supply chain. Enter ECR.

Efficient Consumer Response

In January of 1993, Kurt Salmon Associates, Inc., published the results of its ECR study which was sponsored by four of the major trade organizations in the grocery industry, the UCC, the Food Manufacturers Institute (FMI), the Grocery Manufacturers Association (GMA), and the National Food Brokers Association (NFBA). The results of the study follow.

Efficient consumer response is a business strategy that focuses on improving efficiency of the total grocery supply system. The players are manufacturers, distributors, warehouses, and retailers. The ultimate objective is for manufacturers and distributors to bring better value to the grocery consumer. Efficient consumer response is not a technology, although it employs technologies such as bar-coding and scanning. It represents change; a change in traditional business relationships, moving from the buyer-seller view to the partnership or alliance view; a change in business procedures, built on the adoption of best practices and the reengineering and automation of inefficient processes; and a change in focus, being a responsive, customer-driven system.

The benefits anticipated from the adoption of ECR throughout the grocery chain are reduced total costs, decreased inventories, and improved selection of high-quality and fresh grocery products. Dollarwise, $30 billion can be saved throughout the system and 41 percent less inventory can be stored (from 104 days supply to 61 days). Much of the cost savings will be passed through to the consumer as grocery tends to be a very competitive marketplace. Competition is felt in two ways. One, grocery stores are all fighting for the consumer, who is looking for maximum variety and the lowest price with the highest level of customer service. Two, grocery stores are competing with alternate sources for their products, such as mass merchandisers.

There are four primary strategies of ECR.

1. *Efficient store assortments.* To optimize the amount and variety of products carried based on the profitability of each
2. *Efficient replenishment.* To optimize time and cost throughout the replenishment cycle
3. *Efficient promotion.* To optimize efficiency of promotion throughout the system (manufacturer, wholesaler, retailer)
4. *Efficient product introduction.* To maximize effectiveness and minimize time and cost of product introduction activities

It is anticipated that benefits will accrue for all the organizations involved in ECR as well as for the ultimate consumer. With consumer focus and the sharing of sales information, suppliers can plan their manufacture more accurately and experience fewer out-of-stocks. Suppliers can also improve their relationships with distributors. Distributors can use sales information to improve relationships with both their suppliers and customers. Retailers can expect increased consumer loyalty because they too will have few out-of-stocks and very competitive prices because of cost savings. Finally, the consumer gets increased selection and shopping convenience and the products they want on the shelves, at the lowest prices.

The projection for introduction of ECR has been broken into two phases: the adoption of best practices and full EDI. Best practices focuses on maximizing efficiency and minimizing costs and time during three activities: replenishment between supplier and distributor, replenishment between distributor and retailer, and sales and data collection associated with selling product to the end consumer.

It is believed that when a company works to attain a best-practices business environment, it can expect to get there in 2 years. Since the study was published at the beginning of 1993, the goal was for companies starting this phase of the ECR program at that time to have completed it by the end of 1994. The second phase integrates the individual efforts of the first, to maximize efficiency and minimize manual intensive and inefficient processes throughout the replenishment chain. For the company who has achieved best practices, implementation of full ECR will take another 2 years.

Benefits are expected as follows: two-thirds of anticipated benefits from implementation of best practices, the additional third from implementation of full ECR. Of course, all the changes to systems and processes will require an ongoing investment in dollars and people. It is thought that after the initial 2 to 4 years, savings from inventory reductions and decreased costs will more than cover their ongoing investment. Interestingly, the largest percent of expense to opt for best practices plus ECR is people-related. This includes training at all levels, implementing structural changes in the organization, changing reporting relationships and group and individual job responsibilities, and developing new performance objectives and measurements for business units and individuals.

To get started, companies need to create a climate for change. This requires top-level commitment to the effort. For companies well on their way to comprehensive EDI and ECR programs, the most often mentioned reason for success is total commitment from top management. Likewise, for those struggling to get their program off the ground and gain the interest and buy-in from mid- and lower-level business people, the most often mentioned problem is lack of top management commitment.

When we think of change, we often focus on the move from manual to system tasks, such as reading and interpreting of business documents and substituting on-line interaction for manual key entry. Often this is the easier type of change to enact. Much harder to enact is a change in attitude or relationship. For example, buyers' objectives have been firmly rooted in maximizing the benefits of their own companies at the expense of the seller. In order to do this, they played their position very close to the vest, sharing very little information and gaining advantage at every opportunity. Today, we are asking these same buyers to share information, seek to optimize the benefits of both parties, and sometimes turn over the buying decision to the seller.

For the next step in the process, the company should select two to four partners with whom it wishes to develop an alliance. These should be organizations with whom it already has an excellent business relationship. With each prospective EDI partner, the company should set up a meeting of senior functional representatives from each of the two companies. By broadening the focus from just buyer and seller, the companies can explore ways to improve business in a variety of areas. The outgrowth of this meeting may very well be two or three joint task forces who seek to identify high-payback opportunities for improving efficiency of the business. They may, for example, explore the elimination of invoice deductions if this is a high-cost area today. They may look into the logistics of shipping product, trying to improve truck loading and unloading efficiency. They may explore the benefits of vendor-managed inventory. Based on the results of the task groups' studies, both companies can invest in the implementation of the highest-payback projects, positioning themselves to gain the greatest benefits from their EDI and electronic commerce (EC) investments. They can also develop a long-range implementation plan, phasing in additional EDI and EC applications over time.

Finally, each company must develop an information technology investment program to support implementation of EDI and reengineering of internal business procedures. Figure 8-3 illustrates the EDI activities and business applications you can expect to see in a best-practices ECR implementation. Full ECR implementation would further integrate the efficient and seamless flow of information through the process which, in turn, would further speed up the flow of product.

There are a few noteworthy obstacles to the successful implementation of an all-encompassing program such as ECR. One is organizational and has already been mentioned. That is, without the unwavering support of top management the organization will be unable to break down the barriers that impede progress toward building links with customers, suppliers, and brokers. Another is functional and cultural. Traditionally, companies have organized along functional lines. Each functional area has been structured as a stand-alone unit with its own objectives,

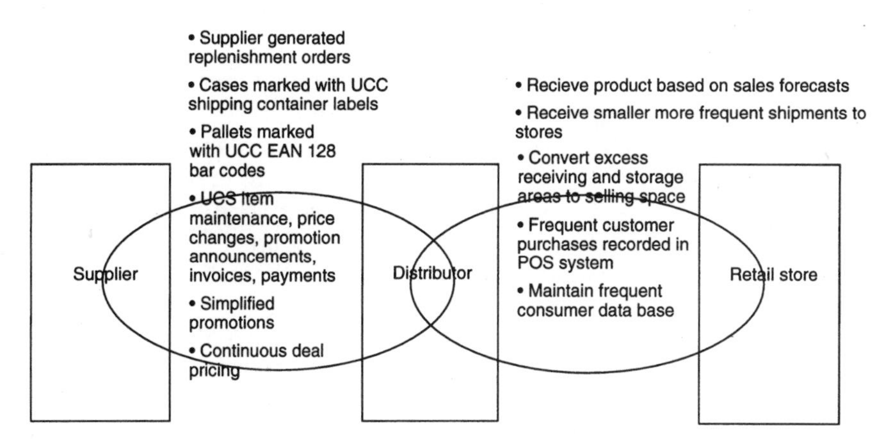

Figure 8-3. EDI and business applications in a best-practices program.

goals, and measurements of success. This structure is a major obstacle for two reasons. One, unless EDI information is disseminated to the various functional areas and personnel that need it, many of the potential benefits of EDI go unrealized. With functionally discrete business areas, there is no one looking out for integration opportunities and the intracompany flow of information between computer applications. As a result, many of the potential benefits attributable to the reuse of EDI data are lost and the information hand-offs between functional areas remain manual and paper-based. Two, in order to optimize the whole, it may be necessary to less than optimize any one segment. With the focus on discrete functional areas, each is attempting to optimize their own position and the big picture cannot help but get lost.

Finally, what is the prognosis for companies in the grocery industry? Kurt Salmon Associates has divided industry companies into three groups. Group 1 contains those who have already or intend to reach best practices by the end of 1994. As early adopters, their cost of implementation will tend to be higher than for those that follow. However, as early adopters, they will tend to gain a competitive advantage resulting in growth rates that will be above industry average. Group 2 contains those that have not yet started to implement best practices but, nevertheless, intend to reach them by the end of 1994. This group will realize much of the benefits of group 1. Their implementation costs will be lower than for group 1. However, they will be playing competitive catch-up with the early adopters. Group 3, those who fail to reach the best-practices goal, will be at a severe competitive disadvantage by 1995. Many will become attractive acquisition candidates. Some will even cease operation by the mid-1990s.

Category Management

To attain the lower pricing objectives mentioned above, companies are beginning to implement category management. Category management furthers the partnership concepts of ECR and is also consumer-focused. In January 1994 the ECR Category Management Best Practices Subcommittee published the executive summary of their soon to be published *Category Management Best Practices Report.* The following description of category management and its components has been derived in large part from the executive summary.

Category management is a process whereby distributors and manufacturers manage categories of products as strategic business entities, so that they deliver better consumer value. It can be thought of as an interactive business process in which business partners codevelop and manage consumer-focused category business plans.

For distributors, category management has resulted in buying-merchandising teams working toward integration of internal procurement, merchandising, and operations business plans and external plans with their suppliers. For suppliers, category management has resulted in the reorganization of supplier profit centers, from individual brand and product focus toward groups of complementary products.

There are six main components of category management. The two core components are

- Strategy
- Business process

The remaining four are enabling or supporting components. They are

- Measuring the category performance (developing the category management scorecard)
- Organizing the company to facilitate carrying out the category management objectives
- Implementing the supporting information technology to facilitate the flow of data
- Developing win-win relationships with trading partners

We'll briefly discuss the best practices of each component of category management.

Strategy. Category management strategy should be an integral part of company policy. However, because category management represents a major departure from the traditional brand management and company strategy, implementing this new strategy must be strongly supported by top management. What's more, category management requires the development of common objectives between consumer, product, and functional organizations within the company.

For the distributor, this means understanding the target consumer; developing new marketing and promotion practices; managing product quality, breadth, and depth; and optimizing functional strategies. For the supplier, this means understanding its consumers and managing product supply, promotion, marketing, and customer strategies. The distributor's goal is to handle each category of products as a strategic business unit and to develop category goals with the end consumer in mind. The supplier develops strategies and procedures that facilitate the accomplishment of distributor category goals.

Business Process. The business process then is a predefined structure of procedures and tasks that supports the accomplishment of category objectives. Category management business process is more concerned with *how* work is done than with *what* actual tasks are being done. The objective is to maximize efficiency of process to produce the highest overall value for the consumer.

Within the business process there are several major activities. They are

- Defining the products that make up the category
- Defining the category objective (purpose)
- Assessing the category
- Establishing the category target rates for its scorecard
- Developing category strategies
- Developing category tactics
- Developing and implementing a category business plan
- Measuring and monitoring the category's progress, modifying established objectives, strategies, and tactics as needed

Category Scorecard. The category management scorecard is a tool used to measure and monitor the progress of the category business plan. Scorecard target rates are codeveloped by distributors and suppliers as part of the category planning process. The scorecard results are derived from traditional financial measures and consumer satisfaction and business process performance figures. Typically the scorecard measures how well the category is fulfilling established consumer-related goals, such as how large the average category purchase is, how many category units are being purchased in one transaction, and how much of the total sales is from this category. It also measures distributor-related objectives, such as total revenue, gross margin, gross profit, inventory levels, and inventory turns. Finally, it measures supplier goals such as revenue, gross profit, supplier share of the category revenue, and overall market share.

Organization. A common result of the implementation of category management is the need for multifunctional objectives and cooperation. Consequently, there is an accompanying growth in multifunctional work teams. Both distributor and supplier need to be organized in such a way as to support intracompany accomplishment of category plan objectives and efficient and effective business procedures.

Information Technology. Category management is a data-driven business process heavy in analysis and statistical measurements. Information technology is a required enabling tool for intracompany and intercompany dissemination of data. Effective category management makes good use of such computer applications as decision-support systems, promotion and pricing models, and store and shelf management software.

Relationships between Trading Partners. Development of more trusting, open, and cooperative relationships has already been a part of EDI and ECR in the grocery industry. However, it is even more important for suppliers, distributors, and retailers to be working in tandem and acting as business partners to successfully accomplish the objectives of category management. Within each company, developing effective interdepartmental teams of professionals puts the right people in place to plan the effort, implement it, and then support it.

We have discussed category management here to show the depth of organizational and business change that accompanies true savings from EDI. Not only are companies increasing their use of technology and enhancing their systems to support new technologies and automate information processing, but they are truly spreading change through every job in every functional area of their organization. The most dramatic of these people changes is evident in the new relationships between trading partners and the requirement to share information and cooperate. In fact, the maximization of the whole instead of the individual parts runs counter to our most deeply rooted emotions and against activities that previously had the highest incentives. Intracompany, this is already an issue, as cross-functional goals and working teams are taking the place of departmental ones. Intercompany, this is even more of an issue.

In fact, often the basic job skills needed day-to-day are not only different but at odds with those that differentiated the "men from the boys" in the pre-EDI environment. This is the reason why top management commitment and support of the effort is the prime requisite of a move into this arena.

Direct Store Delivery

The second segment of the grocery industry is direct store delivery (DSD). Direct store delivery refers to the delivery of product directly to the retailer, where the manufacturer or distributor is responsible for keeping the shelves stocked. In a grocery store, 30 to 40 percent of the merchandise is delivered via direct store delivery. Typically beer, soda, snack foods, bakery items, milk, and general merchandise such as pots, pans, makeup, vitamins, and household implements are delivered in this fashion. In convenience stores, up to 100 percent of merchandise is delivered directly. Direct store delivery is a very labor intensive activity because the vendor delivers product to the store, sometimes counts the remaining inventory, and refills the shelves with product.

As you remember, the A. D. Little study for the grocery industry as a whole had excluded this segment from consideration in its 1980 research because it had been decided that DSD would be treated as a separate issue at a later date. In April 1987, a DSD store-level study was published by A. D. Little. It dealt entirely with the business environment and potential benefits and costs of implementing EDI in the DSD segment of the industry.

The study found that this is a prime area for achieving operational savings for both supplier and retailer. It estimated that with the 500 million deliveries made in this way per year, covering 10 billion line items, retailers could expect to realize $330 million and DSD suppliers $175 million of direct benefits by implementing EDI. Indirect benefits could be even larger, up to 3 times as large for suppliers and potentially very valuable for retailers who could automate their accounting operations and improve line item merchandise management.

This study used as its basis the excellent work begun in early 1985 by the formation of a UCS/DSD task force, whose mandate it was to extend UCS to fulfill the business requirements of DSD. The task force developed two groups of transactions, those to be transmitted between the office of the supplier and the retailer, comprised of contractual data, and those to be transferred from one party to the other at the back door of the retailer, containing information relevant to the delivery itself. The task force realized that in order to process these two groups of transactions, it would need to enhance internal applications to:

- Set up and maintain each trading partner relationship, including the products covered, their prices, and other contractual parameters
- Transact day-to-day business
- Perform billing and accounting reconciliations and authorize payments

The full complement of DSD transactions was labeled DSD/UCS; those dealing with data exchange between offices, NEX/UCS (network exchange of UCS) and those between route driver and back door receiver, DEX/UCS (direct exchange of UCS). All transactions are patterned after the traditional UCS and have been developed with the same architecture, syntax rules, and data dictionary as other UCS transactions. In addition, it was decided that the UPC, being used so successfully in the rest of the industry to identify product to both supplier and retailer, would be employed in this segment in the same way.

Some important findings and predictions of the study are:

- EDI in the DSD segment of the industry can eliminate conflicts and improve productivity, thus eliminating the tension and adversarial relationship between supplier and retailer.
- Proprietary invoice formats and retailer turnaround documents are counterproductive and should be eliminated.
- DEX/UCS will become the predominant means of exchanging data at the store level within the next 10 years.
- DEX/UCS offers significant potential benefits to supermarkets, convenience stores, and DSD suppliers; estimated at 14, 56, and 18 percent, respectively.
- Indirect benefits will be realized through a reduction of retailer manual systems, enhancement of existing business applications, and development of new ones.

A test program was conducted by representative DSD suppliers in 1987. They developed prototype systems and began operational tests in early 1988. As of June, six trading partner pairs were operational. Other standard transactions were developed and went into use in the fall. A report to the industry was completed by A. D. Little in December, 1988.

Let's look at the traditional pre-EDI business environment, systems, manual procedures, and staffing found in this segment of the industry. A retail buyer negotiates with and contracts to buy product from a DSD supplier, sometimes a manufacturer but most likely a distributor. Information relating to their arrangement, such as the products the retailer will carry, the cost to the retailer, and their selling price, is key-entered by a clerk for each store location. That information must be maintained for access by both headquarters and retail stores. Along with each delivery, the supplier brings the invoice. The invoice is used by the retailer at the back door as a receiving document against which it checks in items. All items in the delivery are counted and their prices checked prior to being loaded onto the shelves. If discrepancies are noted between the invoice and actual restocking quantities or between invoice price and preagreed contract price, corrections are handwritten onto the invoice. At night all invoices collected during the day are key-entered by clerks into the system. They are responsible for correctly interpreting the product codes, quantities, and prices on each supplier's proprietary paper invoice form and then correctly entering them into the main office system. Real life experience was that many invoices did not actually make it this far because they were misplaced or damaged during the day at the receiving location. Those that did make it were difficult to read and interpret and were often entered incorrectly. In all, this was a terribly inefficient way of handling deliveries and presented an almost impossible task of collecting and maintaining accurate data.

By contrast, EDI streamlines the delivery process and provides timely and accurate data for both supplier and retailer. Figure 8-4 shows the office-to-office and store-level transactions that would flow between the supplier and retailer using DSD EDI. After setting up the supplier-customer relationship for specified products and at agreed upon prices and delivery schedules, the supplier generates a product information transaction that reflects the terms of the agreement and transmits it to the

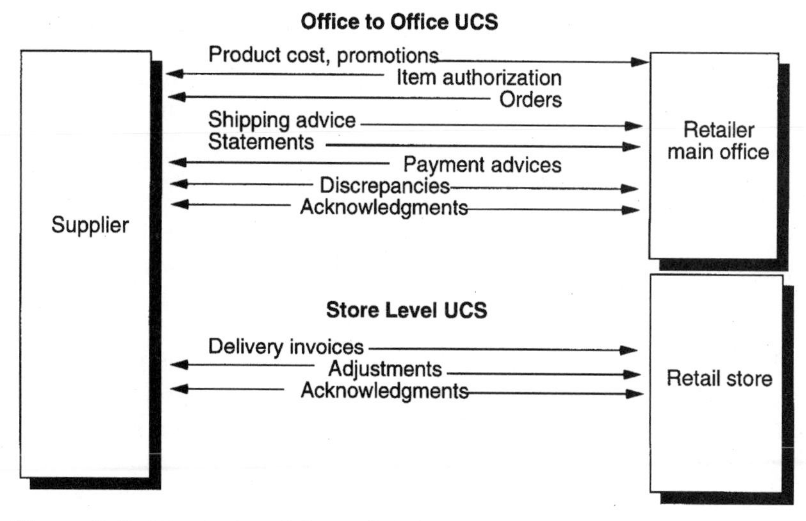

Figure 8-4. Direct store delivery (DSD) EDI.

buyer for his or her perusal and authorization. For each product to be carried from this supplier, the EDI transaction contains the product code, its cost, retail price, and related contract terms. Once the buyer evaluates and agrees to the information, he or she returns an item authorization transaction. All orders, deliveries, and adjustments are verified against this initial information agreed to by both parties. In addition, deliveries are tracked with periodic statements and adjustment invoices, as needed. The retailer may continue to pay by check or may return electronic payment remittances. Both parties transmit discrepancy and adjustment data, as needed. On a day-to-day basis, invoices, adjustments, and acknowledgments of receipt are transferred between the selling agent (route driver) and the buying agent (back door receiver).

There are two ways of effecting this transfer.

1. The route driver arrives at the retailer site with a hand-held device on which he or she has the invoice for today's delivery. That device is then attached to the retailer's system, which displays the product identification (UPC), retail price, and quantity, along with general information about the selling party. Once product has been counted, the invoice can either be adjusted to reflect the actually delivered quantities, or it can be authorized by electronic signature as is. Either way, final invoice information is available to be carried back in the hand-held device by the driver and to be transferred to headquarters by the retailer.

2. The route driver arrives with a "smart card," similar to a floppy computer disk, which again has the invoice on it. This is then slipped into a reader box attached to the retailer's computer and prompts the same display as above. After making necessary adjustments, the corrected and authorized invoice is down-loaded onto the route driver's smart card and is again available to be up-loaded to the retailer's headquarters system.

By having contract information available at the receiving site, the retailer is assured that the vendor is delivering authorized products with retail price correctly marked and has provided the correct price on the invoice. In addition, product identifier (UPC) is available to aid in the scanning process. Product and price exceptions can be found immediately by comparing the invoice information with information on the NEX/UCS headquarters file. Exceptions can be brought up on the screen during delivery and can be dealt with immediately. This eliminates the need for after-the-fact reconciliations and adjustments.

Even special promotion information is available at the receiving location. Prior to running a promotion, the retailer shares its promotion prices with the vendor via EDI. This information is maintained along with the original contract information. When promotional product is delivered, the sale price can be automatically verified by the retailer's system. Some retailers are short-cutting the receiving process even more. On their receiving menu they have options as follows:

1. *Invoice total.* This allows them to just count total cartons, verify their manual count against the vendor's invoice, and authorize or adjust the invoice if necessary.

2. *Item review.* This requires them to count the delivered items as discussed above. This is the more time-consuming method but is the initial method used for a new EDI relationship. However, it is replaced with numbers 1 or 3 when deliveries prove to be accurately reflected in the supplier invoice over a period of time.

3. *Blind entry.* With this option, items are scanned and automatically counted. Information flows directly into the computer system. In this way, the retailer gets the integrity check and count without visually checking and manually counting delivered product. This is the most time and cost-effective delivery and receiving method as it eliminates all data entry and physical counting tasks.

On the vendor side, an automated route-generating system performs two functions. One, it aids in the distribution of product by staging individual store deliveries correctly and on time. Staged orders are then loaded onto the route driver's truck. Two, it generates an invoice for each of the day's deliveries. After making all the daily deliveries, the driver's hand-held device or smart card contains the adjusted and corrected copies of the invoices for the day's deliveries. Upon return to the manufacturer's or distributor's home base in the evening, these invoices are read into the vendor's computer system. Using the daily invoices, periodic statements are computed. These provide the data to perform, after the fact, monthly reconciliations and issue adjustments if needed.

With user-friendly hand-held devices or smart cards and user-friendly menus at the back door of the store, this system is easy to use by both the route driver and retailer who have traditionally been unfamiliar with computers. Other functions can be made available at the receiving site if a microcomputer replaces the dumb terminal hookup to the headquarters mainframe computer. These additional functions would be reflected on the computer screen by additional menu options. With an attached printer, a copy of the invoice could be printed, if necessary.

The three primary benefits of NEX/UCS (sharing of preagreed information between headquarters of the retailer and the supplier) and DEX/UCS (sharing of delivery information), both via DSD EDI, are:

1. To assure product integrity by
   a. Allowing the receiver to verify that the product being delivered has been authorized.
   b. Verifying that the invoice price and retail price match the contracted prices in the retailer's headquarters system.
   c. Identifying the product by UPC code and enabling the receiver to scan incoming product for identification and count.
2. To eliminate the need for after-the-fact key entry of invoice data at the retail store. This, in turn, eliminates transcription errors at both the store and supplier site. With the corrected invoice data downloaded to the route driver's hand-held device or smart card, the route driver returns at night with a correct invoice. When the device or smart card is linked to the supplier's system, the invoices it contains become available to the supplier's application programs.
3. To enhance accounting accuracy by identifying differences between the supplier invoice and the delivery at the time of receipt and by resolving those differences at the back door. In addition, this new method of transacting business eliminates the need to retain and handle paper. As long as each party has access to the data it needs to conduct business, there is no need to retain paper copies.

In order to assure that all parties can trade electronically with all their trading partners, this group initially chose to adopt the same communications standard as the rest of the grocery industry. Each EDI partner purchased a COMM ID from the UCC to identify itself in the electronic transmission and each provided a standard plug connection on its hand-held device, microcomputer, and headquarters mainframe computers. More recently, participating companies have either continued to communicate directly with EDI trading partners or have selected to handle communications through an EDI VAN.

In addition to the above benefits, new business opportunities are anticipated by electronically transmitting data. One such opportunity is to provide the same receiving service in an unattended mode. The route driver would be able to plug the hand-held device into an available connection in order to transfer data to the receiver's computer at the same time that the product is left in a locked receptacle. On unlocking the delivery receptacle, the retailer performs the product and quantity validation. Another opportunity is to include a smart card with the product delivery or to transmit delivery information through an EDI network to support deliveries made by third parties such as UPS. All in all, new logistics methods are possible in the electronic environment which offer timely and accurate data in support of DSD.

It is anticipated that EDI will become the accepted mode of doing business in the DSD segment of the grocery industry within the next 10 years.

The reason for this optimistic opinion is that there are such great benefits to be derived. Those participating in the initial implementations will begin to realize the benefits; those not yet active will begin to experience the competitive disadvantage. With the extensive competition faced by grocery and convenience store chains, these retailers must take full advantage of the cost savings and efficiencies offered by transacting business electronically using NEX/UCS and DEX/UCS data formats.

Recently this author attended a DSD EDI presentation by a major convenience store supplier and convenience store chain. They described their ongoing NEX/UCS and DEX/UCS activity. Amazingly, the potentially automatable parts of the receiving process were still being handled by people. In particular, they had never implemented the automatic system verification of contract product codes and prices. The receiving clerk was still responsible for checking in product. When questioned on this, they reported that they knew of no implementations where systems have been developed to automate the receiving process. What this means is that the potential savings in direct store delivery EDI are not yet being realized. As in most other instances of EDI implementation, the main technology is in place, but traditional business procedures have yet to be replaced by enhanced computer applications.

SuperValu

SuperValu is an example of a major grocery retailer implementing EDI throughout the company. The main reason, according to George Klima, former head of the SuperValu program, was to eliminate information float. From the data received via EDI, SuperValu business applications provide to the functional areas all the information they need to transact business. In addition, by making information accessible to the various functional business areas, they have begun to integrate their business functions.

SuperValu was one of the leaders in support of the UCS standard for the grocery industry. With top management support, it embarked on an aggressive program to incorporate EDI into all areas of the company and to eliminate paper transactions for both intercompany and intracompany messaging. In order to do this, it conscientiously evaluated data requirements for transacting business and then redesigned internal applications to support the information needs of its business and business people. It has developed a process flow where human intervention is used only for exception processing. All in all, the company believes that the implementation of EDI did not really cost them anything because its cost was balanced by the savings they realized when several applications that were to be developed were eliminated.

Let's take a look at the cost savings that SuperValu is realizing and/or predicting by trading electronically in various transaction sets.

- With the 6500 purchase orders it generates per week, it is saving $1.20 to $1.30 per order.
- By ordering from EDI-capable vendors, it can reduce order lead time by one day.
- It predicted a 15-person reduction of buyers by March 1989.
- It expected to save $6000 per day in purchase order–receiving–invoice reconciliation costs by the same date.

In order to achieve these goals, SuperValu developed and continues to develop new applications that use electronically received data more efficiently. In addition, it moved aggressively to increase both its number of EDI trading partners and the number of different electronic transaction sets that it transmits and receives. By March 1989, it hoped to be processing and paying 80 percent of its invoices electronically. Today, SuperValu reports 1200 EDI partners with whom it trades directly and an additional 1300 with whom it trades electronically through brokers.

Following are some of the applications contained in its EDI and business systems:

- *Request for inventory count.* Used to get an accurate count of product on hand, and as input to the invoice reconciliation process.
- *Receiving against invoice.* Reconciliation of the invoice count, the open purchase order, and the quantity being invoiced.
- *Computer generation of charge back.* Developing a debit transaction to reflect reconciliation errors.
- *Promotional announcement processing.* In this system, the vendor or broker maintains SuperValu files with the buyer's approval. This ensures that manufacturer promotions are taken advantage of by SuperValu. The computer program recognizes a promotion opportunity, places the order, and then calculates and passes on allowances to the individual stores. Currently, SuperValu is handling promotional announcements via EDI. This application represents savings for SuperValu in the *million dollar range,* as many promotions were missed prior to its implementation.
- *Price changes.* SuperValu gets about 20 price changes per week. With the new system, which receives notification of price changes electronically and then process them automatically, benefits will be:
  - Receipt of accurate and timely price change data

  - Improved customer service
  - Improved buyer-seller relationship
  - Maintenance of accurate SuperValu files by price bracket
- *Item maintenance.* A system in which the vendor maintains the item UPC plus other alias codes in its computer for the customer. Benefits of this system are timeliness and accuracy, as above.

SuperValu realizes that it will receive maximum benefits from EDI only by increasing its volumes and the variety of electronic messages that it supports. Because of this, it maintains very aggressive goals. Its integration of EDI data into daily processing routines has been a strong example to others in the grocery industry.

General Merchandise

This second half of the retail discussion will focus on the sale of general merchandise through a mass merchandiser. We'll take an in-depth look at the implementation of EDI by Service Merchandise, one of the largest mass merchandisers because of its organized and well-planned approach to implementation. Then, we'll focus on QR systems: what they are, who is doing them, and what the players are realizing in the way of benefits. Finally, we'll take a look at a few successful QR programs.

Mass Merchandisers

For purposes of this discussion, a mass merchandiser is defined as a low-level retailer carrying a large variety of merchandise. Such a store competes for consumers mainly on the basis of price and customer service as it and all its competitors tend to carry the same products. Demand for product in this type of store is inelastic, meaning that a small change in price brings about a large change in demand in the opposite direction; i.e., a small price *increase* results in a large *decrease* in demand. This type of retailer must look for ways of decreasing internal costs in order to provide product at the lowest cost while earning a reasonable net profit.

As mass merchandisers are notoriously large purchasers, it is natural for them to attempt to increase efficiency in their inventory management activities. As a first step, they normally centralize their purchasing areas. With all purchasing done through one department, they are already able to realize substantial benefits over a distributed environment, no matter the medium; paper, telephone, or electronic. By centralizing purchasing they:

- Devote only one staff of buyers to the purchasing function and order from their suppliers as one major customer for all store locations
- Control inventory for all locations out of a single system
- Get substantially lower costs and better delivery schedules for their large quantity purchases

Of course, in order to purchase from a central system, the mass merchandiser must have appropriate systems in place in its various stores to collect sales, receiving, and return data; transfer that data to the central location; make all the necessary decisions at the central locations; and send updated information back to the store locations.

This author has used Service Merchandise as a representative mass merchandiser because of its early and aggressive implementation of EDI. While others today have equaled or surpassed the Service Merchandise program, it will be described in some detail because it is so organized and structured in its approach. At the end of the description of the original program, an update of its and other mass merchandisers EDI activity today is included.

Service Merchandise

Service Merchandise, one of the pioneers in EDI, supports the ANSI X12 standard, does all its purchasing out of a central location, and is on-line to its various stores daily. Associated with each sale, sales information is collected at the store via a point-of-sale system. Likewise, product receipt and return information are also collected at the store site. Each night store transactions are up-loaded to the main Service Merchandise computer and are used to generate chain-level purchase orders to the various vendors. Then, files with current retail prices and updated inventory figures are generated and down-loaded to the individual stores along with copies of the orders for that store. These orders are compared to product deliveries during the day. When delivered products and quantities vary from the orders, the system creates variance reports. Service Merchandise is able to realize the benefits of a centralized purchasing system and store-level information systems because both handle data in electronic format. The company not only takes advantage of timely and accurate information but has also eliminated manual intervention and mail float. In addition to streamlining the purchase of product, Service Merchandise has also sought to reduce the effort it previously expended to handle and reconcile incoming invoices. Not only was this procedure time-consuming and expensive, it was also fraught with errors. To automate the reconciliation procedure, Service Merchandise developed a computerized matching system that compares the open purchase order sent via EDI, receiving information, and the invoice received via EDI. These are just some of the ways in which Service Merchandise has used EDI as a vehicle to streamline business between itself and its vendors. For this effort, the EDI standard it chose to support was ANSI X12. Unlike the grocery industry, mass merchandisers support this standard for intercompany trade of general merchandise. For grocery products, they may support UCS.

Service Merchandise was one of the first mass merchandisers to develop a full-fledged EDI program. In 1985, directives came from the chairperson of the board to the IS department to design and develop the systems necessary to support the electronic purchase order and invoice. Its purchasing and accounts payable users became active after the fact.

Service Merchandise used the tremendous amount of clout it had with its vendors to encourage them to implement EDI. The initial letter it sent to the vendor community, signed by the vice president of merchandising, was fairly mild, relating Service Merchandise's desire to use EDI. The second letter gave dates by which the vendor was expected to be up and running. The top 200 vendors, measured by amount of business, were given the choice of transmitting directly with Service Merchandise or through an EDI VAN. One hundred and fifty of them chose to transmit directly, causing Service Merchandise to develop its own network to handle EDI transmissions. It chose to support the 2780/3780 bisynchronous protocol for communication access, since most of its trading partners could already support it. Vendors below the top 200 were directed to transmit their EDI data through McDonnell Douglas, later known as BT Tymnet and today owned by MCI, Service Merchandise's endorsed VAN.

McDonnell, in turn, developed a trading partner marketing program to help get Service Merchandise's trading partners on line. It sponsored trading partner meetings which were attended by Service Merchandise representatives and those of prospective EDI trading partners. Between the pressure exerted by Service Merchandise and the efforts of McDonnell, many targeted trading partners implemented EDI.

However, over time the trading partner program fell into disuse and the pace slackened considerably. Even though the number of EDI partners grew substantially, Service Merchandise was disappointed with the ongoing results of its trading partner program and chose to endorse a second VAN who would step up the marketing of EDI to potential EDI partners.

One of the things that became apparent as the Service Merchandise MIS staff got ready for electronic trading was that a good deal of change was necessary to make its computer applications EDI-capable. For one thing, it had to automate all manually performed edits and have errors kicked out as exceptions. For another, it needed to route exceptions to the appropriate people and program the system to alert them of the problem. Then, it had to find ways to codify its data fields. In the past, incoming narrative on business documents was read and interpreted. Outgoing messages were printed and read and interpreted by the receiver. In the EDI environment, all data must be understandable to a computer program and be acted upon automatically. Finally, it had to integrate the "gateway" computer application, the application receiving EDI data, with subsequent business applications in order to seamlessly route incoming EDI data and retain copies of incoming EDI transaction sets for use at a later time in matching and reconciliation procedures.

As systems were developed for the EDI environment, the underlying philosophy was to allow correct transactions to flow automatically through the system with no manual intervention and to kick out error conditions in exception reports and pass them to the appropriate staff. Has EDI been successful for Service Merchandise? Yes! As a result of EDI and its automated invoice matching system, it has reported a reduction in its accounts payable staff and effort of 80 percent. As a matter of fact, based on Service Merchandise's aggressive EDI plans and its planned integration of EDI into several facets of its business, it took a very organized approach to the EDI program. Figures 8-5 through 8-7 contain lists of the tasks it performed during the planning, organizing, and implementation stages of EDI. Following are discussions of each task.

Planning for EDI

1. *Identify and prioritize your objectives.* In the case of Service Merchandise, the objectives were to:

1. Identify and prioritize objectives.
2. Define the scope of the project.
3. Select trading partners.
4. Choose a standard to support.
5. Develop a plan for communicating with trading partners.
6. Develop a plan for handling translations to and from your internal application.
7. Develop a realistic budget for the EDI project.
8. Secure management support.

Figure 8-5. Service Merchandise—tasks performed during the planning stage of EDI.

- Reduce lead time
- Reduce internal costs
- Reduce the error rates associated with ordering and receiving
- Eliminate manual key-entry errors

Service Merchandise understood that these objectives would be realized over time and that integration of applications within the company made additional indirect benefits possible.

2. *Define the scope of the project.* Service Merchandise decided to handle EDI as a central capability, buffering its business systems from the EDI system. It chose this as the path of least resistance in the long term since adding new EDI transaction sets would require developing new links to its application programs. This path also turns out to be the least expensive because it requires only one EDI translator, capable of translating all transactions in and out of the various versions of the EDI standard. This same centralized translator can be utilized by all internal departments and divisions of the company.

In defining the scope of the EDI effort, Service Merchandise also included evaluation of current business procedures and the changes that would be needed to support EDI. As far as personnel was concerned, Service Merchandise planned on a teamwork approach with its targeted partners.

3. *Selecting trading partners.* Service Merchandise selected pilot partners early, choosing a small number from those with whom it did a high volume of business. By selecting large-volume trading partners, it maintained high visibility for the EDI program and discovered the vast majority of system bugs with the first few partners. After the initial pilot, Service Merchandise wanted to quickly increase its EDI trading partner base. In anticipation of the end of the pilot period, it selected prospective EDI vendors by targeting its top 600 trading partners.

4. *Choose an EDI standard to support.* Service Merchandise investigated the usage of EDI standards within its industry and especially those supported by its trading partners. It participated in industry groups and took an active part in the standards committee in order to assure that its data and business needs were met by the EDI standard.

5. *Develop a plan for EDI communications with trading partners.* Service Merchandise offered its top 200 vendors the choice of communicating either directly or through an EDI VAN. All other partners were expected to communicate through a VAN to eliminate communication support for smaller and less-sophisticated vendors.

6. *Develop a plan for handling translations to and from internal applications.* Service Merchandise is totally committed to the ANSI X12 public data standard. It purchased an EDI translator instead of developing a translator itself.

7. *Develop a realistic budget for the EDI project.* In order to effect a successful implementation of EDI, Service Merchandise realized that it would have to spend time and money on:

- Educating user and data-processing staff
- Traveling to standard committee, industry trade group, and trading partner meetings
- Designing, programming, and testing applications systems that support its EDI
- Developing documentation for internal use as well as for trading partners

Since the project would have floundered from lack of resources without budgeting these expenditures in advance, Service Merchandise requested a development budget and received approval for it.

8. *Secure management support.* In the case of Service Merchandise, the CEO led the bandwagon for EDI. There was never a lack of top management support. However, those responsible for implementing EDI were aware that they had to get the support of management at all levels, especially the direct involvement of the account representative of each trading partner they targeted for conversion to EDI. Without user involvement, development of new business procedures for the electronic environment would have been impossible. And without user support and advice, they would have antagonized just those staff members who were needed to ensure a successful implementation.

With This Plan in Place, Let's Look at the Service Merchandise Plan of Attack for Organizing the Program

1. *It dedicated resources to the project.* It selected a person to attend committee meetings and assigned technical people to assist the business users to define their requirements. It then designed system enhancements, programmed them, and installed them.

2. *It put into place a program to approach trading partners.* It conducted seminars, visited trading partner locations, and sent out mailings. It also enlisted the aid of its EDI VAN to cosponsor trading partner marketing seminars.

3. *It selected two to three pilot partners.* It used high volume of business, high interest in EDI, and current EDI activity as criteria for its selection of EDI partners.

4. *It identified its EDI data requirements from the paper documents it used to transact business.* It found that a good deal of data traditionally

1. Dedicate resources to the project.
2. Put program in place to approach trading partners.
3. Select two to three pilot partners.
4. Evaluate current paper transactions.
5. Establish prototype pilot procedures.
6. Develop clear and easy to understand documentation.

Figure 8-6. Service Merchandise—tasks performed during the organization stage of EDI.

included on paper is redundant or extraneous in EDI. In developing data requirements for its EDI transactions, it eliminated redundancies and pared the purchase order transaction set to the 300 to 350 characters absolutely needed by a supplier to process the order. In this way it built in cost savings in transmission charges.

5. *It established procedures for its pilot program.* These included testing each new EDI partner's communications and data transmissions, running and comparing parallel EDI and paper systems, and migrating successfully tested partners from test mode to production. By developing the prototype procedures up front, it was able to quickly bring up trading partners and fulfill its goal of bringing up several hundred vendors.

6. *It developed clear and easy to understand documentation.* This documentation explained the data requirements, communication requirements, and support procedures. It was made available to prospective EDI partners to facilitate their own EDI implementations. Finally, an implementation program was developed to encourage trading partners to trade electronically with the company. While Service Merchandise could exert a great deal of influence over its vendors' actions, it realized that success in this venture was dependent on the type and amount of support it offered to its vendors.

Toward This Implementation Goal It Performed the Following Implementation Tasks

1. *Developed a sales program to "sell" EDI concepts to its trading partners.* It developed a presentation that visually explained how EDI works, what the hardware and software requirements are, and what benefits vendors could expect to realize from electronic trade with Service Merchandise.

1. Develop sales program to promote EDI to trading partners.
2. Help targeted trading partners to develop cost-effective EDI program.
3. Encourage trading partners to commit to a time frame.
4. Offer Service Merchandise staff as consultants.
5. Invite trading partners to marketing seminars.
6. Get top management to participate in EDI project.
7. Offer incentives to trading partners.
8. Follow-up! Follow-up! Follow-up!

Figure 8-7. Service Merchandise—tasks performed during the implementation stage of EDI.

2. *It helped its targeted trading partners to put together a cost-effective system for communication and translation* that would utilize the hardware they already had.

3. *It encouraged its trading partners to commit to an implementation time frame* so that Service Merchandise could schedule trading partner implementations.

4. *It offered its own staff to provide assistance to trading partners,* as needed.

5. *It invited selected trading partners to attend marketing seminars.* Realizing that the Service Merchandise buyer has leverage with the vendor's salesperson, letters to vendors were directed from the Service Merchandise account representative and were signed by the vice president of purchasing on behalf of the buyer.

6. *It got top levels of management to take part in the project.* Its CEO was very vocal on the matter of EDI, both within the company to staff and outside to trading partners. This further showed the strongest possible corporate commitment to the goal of trading electronically with suppliers.

7. *It offered incentives to targeted trading partners to induce them to sign up.* Even though the supplier would realize benefits by implementing EDI, Service Merchandise sweetened the pot by offering a favorable vendor status to those who signed up to trade electronically.

8. *Finally, follow-up.* As a result of the initial EDI letter, invitations to meetings, and the meetings themselves, only 10 percent of Service Merchandise's targeted suppliers signed up without additional follow-up.

All others got follow-up material and sometimes on-site visits before they decided to implement EDI. Service Merchandise realized that if it did not give the extra effort, it would fall short of its goals.

Using this organized approach, Service Merchandise targeted its top 600 suppliers for EDI implementation. It selected the purchase order as the first electronic transaction with the invoice following closely on its heels. Service Merchandise's objective was to encourage its trading partners to adhere to a short time frame in implementing both transactions.

It initiated its program in 1985. Since then, it has brought up close to 615 vendors, which represents 80 percent of its top vendors. However, this represents only about 33 percent of its total vendor base of almost 1900. Today, Service Merchandise sends 90 percent of its orders via EDI and receives 80 percent of its invoices in the same way. Of those that use EDI, 122 continue to transact business directly with Service Merchandise. All others have chosen to use an EDI VAN as communications intermediary. The greater than 1200 vendors not yet doing EDI are proving to be difficult to convert. They are typically the smaller vendors to Service Merchandise and sometimes are smaller companies. To reduce the manual processing of orders and invoices, Service Merchandise uses an outside service that converts EDI orders going to non-EDI vendors to a print format and faxes or mails them. The service also converts incoming invoices from print to EDI and transmits them to Service Merchandise.

While Service Merchandise has realized substantial benefits by sending the electronic purchase order to its vendors and even greater benefits by eliminating staff and paper processing in the accounts payable area, it has not moved on to Quick Response (QR) and increased the number of different transactions that it uses, as so many of the retailers have. Service Merchandise believes that this is because it originally sold EDI internally to top management and business users as a way to eliminate staff and reduce paper instead of as a tool to improve the movement of timely data and to effect inventory management. Even though it has over 600 trading partners, it considers itself to be somewhat stalled in a long-range approach to EDI and electronic commerce.

Quick Response

Just as ECR is a business strategy focusing on improving efficiency throughout the grocery supply chain, QR is a business strategy seeking the same goals for the retail (general merchandise) industry. And just like ECR, QR is a combination of the use of various technologies such as bar-coding, scanning, customized ticketing, and automated carton handling, and business practices that enable the swift and accurate movement of product through the replenishment cycle. The players in quick response are the manufacturers, distributors, and retailers. The benefits accrue to all. The big winners are consumers, who have access to the products they want, when they want them, at the best possible prices.

Quick response was originally conceived as a means to respond to the intense competition from overseas manufacturers that vendors were experiencing as early as the 1970s. Previous attempts to gain competitive advantage included seeking government legislation to protect domestic firms from overseas competitors and investing heavily in equipment to bolster productivity of domestic firms. This resulted in highly automated fiber and fabric manufacturing firms with the highest rate of productivity found in US manufacturing companies, Finally, there was a "Crafted with Pride in the USA" campaign aimed at raising consumer awareness of the country of origin of the goods that they purchase and encouraging them to purchase US-manufactured products.

In 1985 to 1986, Kurt Salmon Associates conducted an analysis of the supply chain in the retail industry that showed that while discrete segments of the chain were efficient, the overall efficiency of the whole was very low. It identified ways in which individual players in the supply chain were actually adding significant costs to the overall flow of product. The study estimated that the dollar loss to the system was $25 billion; two-thirds of which was due to mark-down losses by the retailer or manufacturer and to sales lost from out-of-stocks by the retailer.

Again, similar to ECR, QR is a partnership strategy in which retailer and supplier work together to respond more quickly to consumer needs by sharing point-of-sale (POS) information and by jointly forecasting future demand for replenishment of product. Using QR, a 75 percent reduction in lead time is not unusual. Not only is product received sooner in a QR environment, but the management of inventory throughout the supply chain is of paramount importance. In some cases, the vendor even manages the customer's inventory, eliminating the need of the customer-generated order. This service is called *vendor-managed inventory* and is covered below in the discussion of the apparel sector of the retail industry. Figure 8-8 contains a list of the primary EDI transactions traded between a manufacturer and retailer in a QR program.

One initial barrier to implementation of QR was the lack of an industrywide, agreed-upon EDI standard for product identification. In 1986, the Voluntary Interindustry Communications Standard (VICS) committee developed EDI standards for the retail industry by defining the subset of the ANSI X12 standard that provided the information needed to transact business in the retail industry.

Because of this activity, VICS is often credited with accelerating the EDI process for the retailer and its suppliers. VICS was formed because many large companies felt that the existing standards associations were

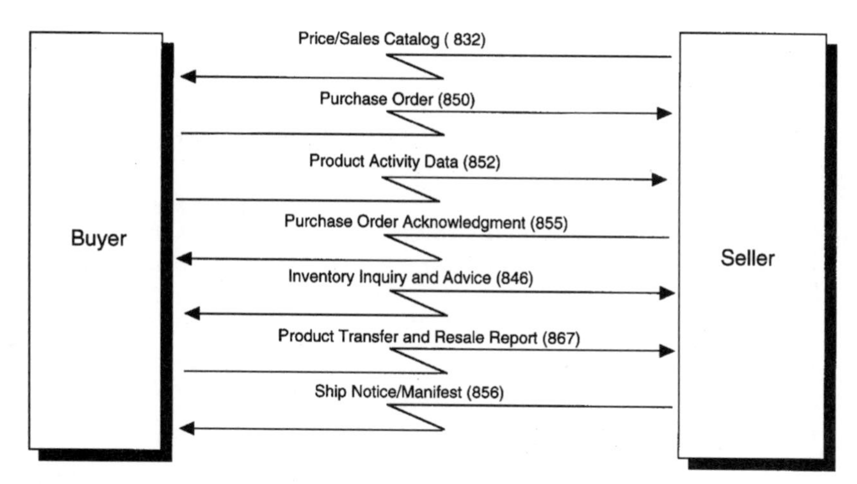

Figure 8-8. EDI transactions traded between manufacturer and retailer in a QR program.

not directly addressing the needs of the apparel segment of the retail industry and were not effectively supporting the implementation of related new technologies. Member companies of VICS include manufacturers, textile mills, and retailers. This organization used the UCC, with its implemented base of grocery companies, as its organizational model. While they chose to support the ANSI X12 standard instead of the UCS standard supported by grocery companies, they did decide to use the Uniform Product Code–Version A (UPC-A) managed by the UCC as the methodology for identifying items in this industry. Initially, the leaders in the industry worked through VICS to set up prototype guidelines of standard usage for the industry. The full membership evaluated these guidelines and approved a revised version for use by all. VICS published an implementation guide that contains guidelines for use of the ANSI X12 standards for the retail industry. It continues to update the guidelines on a semiannual basis.

The result of defining the guidelines for usage of the VICS subset of the X12 standard has been the rapidly expanding use of EDI in the retail industry. As a matter of fact, this industry has far surpassed the use of EDI in the grocery industry; most notably for the ASN transaction set. The success of QR in the retail industry can be attributed to the vision of a few business leaders who realized that the ability of the entire industry to compete in the world marketplace can be vastly improved if suppliers and customers work in concert to eliminate inefficiencies and waste throughout the replenishment cycle. They also realized that technology, while supporting the overall effort, is not the solution in itself. The main effort would be to develop new business procedures and relationships and to change corporate goals and existing corporate cultures.

Quick response became the de facto definition of the highest in customer service in the apparel sector of the retail industry. Below are descriptions of some apparel QR programs.

Apparel Sector of the Retail Industry. In looking at this sector of the retail industry, we see a continuum of establishments which include:

- The *mass merchandiser* who competes almost entirely on price.
- The *value-oriented department store,* competing predominantly on price, but with some sensitivity to fashion.
- The *mainline department store,* a blend of price and fashion sensitivity.
- The *high-profile department store,* competing almost entirely on fashion, with little sensitivity to price.
- The *specialty store,* focusing on a particular market niche. This type of establishment may be anywhere on the price-sensitivity scale, depending on the amount of competition in the market niche and the quality of merchandise that it carries.

As we might expect, there is a direct relationship between an establishment's price sensitivity and its use of new technology to streamline procedures and cut internal costs. Examples of new technologies being employed in this industry group are bar-coding, scanning devices, point-of-sale (POS) systems, and EDI.

Let's look at the general economic conditions that exist in the retail industry today. The consumer has an increased amount of disposable dollars; however, the value of the dollar has decreased, even with low inflation rates in recent years. Cost of goods is increasing, especially for those firms employing American labor. In addition, we continue to have too many stores. With population shifts out of the major cities, shopping centers have sprouted up ad infinitum in suburban areas all over the country. Each retail location is expected to carry a full line of inventory to cater to the local population. Because of the proliferation of store locations, coupled with the requirement for a large variety of product choices, earnings are depressed. Add to this the number of mergers and acquisitions that have been occurring in recent years, and you get a very uneasy business picture, comprised of companies desperately looking for the means to increase efficiencies and decrease costs.

While mergers and acquisitions may ultimately provide an environment in which companies can realize benefits of scale, in the shorter term this same situation can delay the use of new technology by providing a distributed and varied computer, system, and procedural environment. Even though EDI is most efficient and cost-effective when implemented as a central corporate function, that may not be possible in a recently merged corporation.

One of the main business problems facing retailers is a lack of control over their inventory. Consequently, they see inventory management as one way to vastly improve their business situation. In the past, there was noticeable lack of control over inventory. This was because buying decisions could not be based directly on actual sales activity, as sales were rarely tracked as they occurred. Also, in the past the retailer was willing to receive a predetermined size and color mix of items for a style that it carried. Today the retailer is ordering in direct response to demand and vastly reducing overstock and understock situations. This, in turn, has eliminated both drastic markdowns of surplus product and loss of sales from empty shelves.

In the past, retailers were left in the dark as to what they were actually selling. Inventory was not tracked to the lowest level, e.g., size and color. Also, quantities of product were not differentiated as to their inventory category such as sold, available for sale, and in transit. With the adoption of the UPC number and its acceptance by both retailer and manufacturer to identify product down to the style, color, and size level, both retailers and manufacturers can easily identify and track product. In many cases, acceptance of the UPC number has eliminated the need for each partner to maintain product aliases and cross-reference to its own product-numbering scheme. It has also provided the vehicle by which a POS system can track items down to the specific selling unit. Use of the bar-coded representation of the UPC number has also been adopted by many companies. Using the bar-code symbology allows retailers to scan bar-coded carton and item labels and automatically up- or downgrade the appropriate inventory category. Retailers scan the bar-coded label when receiving the product and again at the time of sale using their POS systems. During the former, they increase their inventory-on-hand figure; during the latter, they reduce inventory on hand and increase inventory sold. By collecting sales information both retailers and suppliers can develop tracking reports that detect buying trends and alert buyers and suppliers of fast- and slow-moving items. It is just this type of information that helps retailers to make smarter buying decisions and allows manufacturers to more closely match replenishment to actual demand.

While the UPC code provides the vehicle to universally identify product, it is not a full solution. Both retailers and manufacturers need smarter business systems to make more timely and accurate decisions. In addition, both need to close the gaps between discrete business systems, to allow the uninterrupted flow of information throughout their organizations. Also, with the introduction of a more sophisticated POS system, the retailer can gain more accurate demographic information about consumers and develop promotions that will net higher revenues and profits. Consequently, there has been a trend for POS software developers to enhance their products with additional functionality such as automatic use of scanned product data for various new purposes, price lookup capability, and development of sales-tracking reports from current and historical data.

Retailers have also sought to improve their receiving procedures and tracking of inventory quantities. Here, manufacturers are being asked to assist the process by marking each shipping carton with a bar code that uniquely identifies it. This shipping container marking is scanned in the retailer's distribution center to automatically send the carton toward the correct store-destined delivery truck. With increased automation in the receiving area, shipments received from a manufacturer can be cross-docked to the store delivery truck by means of a completely automated system. It is not unusual for a distribution center to receive product and get it on its way to the store within minutes. When the retailer has received an advance shipment notice from the manufacturer, identifying the product contained in each shipping carton, the retailer can upgrade inventory counts of products received merely by scanning the carton bar code. The need for the retailer to scan or count individual items in the carton is eliminated. Also eliminated is the time lag between receipt of goods and their placement on the selling floor. To further reduce lag time during the receiving process at the store location, manufacturers are being asked to provide a customized label to each item containing the UPC, retail price, and other retailer-requested information. More recently, manufacturers have been asked to further facilitate store restocking by folding product in such a way that it fits immediately on the store shelves or by shipping goods on hangers as requested by the retailer.

Let's take a look at how two innovative apparel manufacturers and a specialty store chain have integrated EDI in their businesses.

Levi Strauss & Company

Prior to its support of the ANSI X12 standard , Levi had designed a proprietary standard for electronic communication which it used with a number of its large customers and suppliers. However, after the development of ANSI X12, Levi became a strong advocate of its use instead. It has held leadership roles in various X12 subcommittees and task groups as well as on the VICS committee.

Levi receives 58 percent of its orders measured by dollar volume via EDI. Some of its large customers began doing EDI by sending purchase orders in their own proprietary format. However, over time Levi was able to migrate those partners to the X12 standard. Today, Levi supports only the X12 standard for all EDI transmissions. It, like most other apparel manufacturers, has adopted the UPC number to identify its manufactured items.

Today, Levi offers QR for its major customers. By developing automated or semi-automated systems to support the entire business cycle from capturing POS information to replenishment of inventory, Levi has become ultimately responsive to its customers' needs. In addition, Levi conducts vendor-managed inventory (VMI) programs with some retailers, where it manages inventory replenishment on behalf of the customer. It not only has devoted internal support to its own multiple EDI-related business applications, it also assists retailers to implement automated systems which track sales, handle receiving information, manage inventory, and order correctly.

Several years ago, Levi surveyed its customers for interest in EDI. It found the response was even higher than expected in favor of automated systems and electronic transmissions. Because customers in general did not have systems in place to enable them to automate the order of replenishment stock, Levi used the favorable response it received to justify development of its own model stock management system. Upon its completion, this system became the cornerstone of its VMI program. When setting up a VMI program with a retailer, Levi assists the retailer to design model stock requirements for Levi's products. In order to develop the best model stock for the retailer, Levi shares with the retailer results of its analyses of consumer purchasing patterns. The items (styles) that the retailer wishes to carry, along with their size and color mix, are entered into the Levi system as the retailer's model stock product requirements. On a day-to-day basis, the retailer collects and transmits to Levi via EDI either POS data for sales occurring since the last transmission of inventory-on-hand counts. Levi compares these figures to the model stock defined by the retailer, computes suggested replenishment quantities, and either returns a suggested purchase order to the buyer or actually develops an order itself and reports shipment quantities back to the retailer. When a retailer signs on for the Levi VMI service, it is assured of always having on its shelves the inventory it needs to satisfy customer demand for the entire Levi line of apparel.

To further facilitate product replenishment at the retail store, Levi transmits an ASN to the retailer that provides details of the product mix in an impending shipment and relates its expected arrival time. Included in this transaction are the serially generated numbers that uniquely identify each of the shipping cartons and the detailed description of contents in each. The identifying serial number is displayed on the outside of each shipping container using the industry standard 128 bar-code symbology. Using the ASN, retailers can prepare shelf space for incoming goods and allocate staff to unpack the cartons and restock the shelves. By providing carton contents in the ASN, the retailer can upgrade its inventory counts without having to physically count the individual items contained in each carton. Some retailers have begun to use the ASN as a receiving record. Merely by scanning the bar-coded label on the shipping carton, they are able to verify receipt of the contents described in the ASN transaction set. Additionally, Levi tickets each item with a customized tag containing the bar-coded UPC number and retail price of the item. This eliminates the need for the retailer to mark each item prior to stocking the shelves and sets the stage for automated scanning or key-entering of the UPC into a POS system at the register.

In some cases the ASN eliminates the need for the invoice since it already contains all the same data as the invoice except price. When prices have been agreed to prior to the sale, the invoice becomes redundant to generate a payment. This may be an attractive arrangement for retailers because it eliminates the need for them to develop an automated invoice reconciliation system. When invoices are still used, EDI allows computerized reconciliation of the invoice down to the line-item level, an activity that is often impractical with paper business documents.

As for EFT, Levi relates that there is little activity in this area. Retailers will need to believe that they will realize benefits before they will convert to payment via EFT. The main retailer concern seems to be loss of float. Because of this, it may be of value to both parties to develop float-neutralizing terms to encourage EFT payments.

Today Levi is considered a high-service–level supplier. It has implemented a large variety of EDI transactions that support day-to-day business as well as maintain the currency of product information. In addition, it offers a flexible package of value-added services that provide the retailer with the ability to order, receive, and restock in the most efficient and cost-effective manner.

KG Mens Stores

KG Mens Stores, a chain of 115 specialty stores carrying men's clothing, is a showcase customer, having implemented EDI and taken advantage of the value-added services that Levi provides.

About 8 years ago, it took a close look at the industry and at its economic situation. It realized that it needed to reduce safety stock and inventory and cut internal costs because profit margin and turnover of inventory were both too low. Prior to double-digit inflation, it had not perceived inventory level as a major problem. It also realized that competition was growing, partially from other specialty chains, but mostly from the traditional department stores beefing up their specialty departments. So, KG was not only interested in carrying less inventory but in carrying the right inventory. Its buying traditionally had been done by lot. The typical apparel lot contains a predefined mix of size and color within a style. It was sometimes unable to fill a customer's demand for a popular size or color using this buying technique. At the same time, KG was left with unpopular sizes and colors which it had to offer at large markdowns.

Another reason to try to improve buying was that shipping costs were escalating. KG often needed to transfer stock between its own retail outlets. It wanted to buy right for each store in order to minimize both shipping costs from the manufacturer and product transfers between stores. Fortunately, purchasing at KG was a centralized function, so it was well positioned to gain major benefits from streamlining the purchasing function. It regularly purchased merchandise from about 500 vendors.

KG's main philosophy was to use new technology to reduce costs, thereby increasing bottom-line profits. Because it perceived that savings would be highest by automating the purchasing area, it started there. It contacted its vendors to find out if they were interested in trading electronically. For a short time, it considered developing its own EDI standard but decided instead to support the already developed ANSI X12 standard. It became an active member of VICS, which allowed it input into the standards process. Internally, KG developed business systems for its mainframe computer that supported the VICS/ANSI X12 guidelines.

Today, its purchasing application is almost completely automated. Sales information is accumulated at the store level, tracking UPCs which specify style, size, and color. Nightly, sales information is pulled from the stores to the central computer. Ninety percent of orders are then generated automatically through an automated replenishment system. This replenishment system uses actual sales figures to determine order levels of products regularly stocked in the stores. It uses trends in demand of similar items for new products. In addition, it uses seasonal adjustments and other parameters to further tailor orders to actual consumer demand.

Store-level responsibilities at KG include customer service, morale of employees, unloading of shipments, and display of goods on the store shelves and racks. While ordering is done at the corporate level, receipt of goods is done at the stores. One hundred percent of its stock is drop-shipped to the individual store locations. KG has no warehouses for storage of excess inventory. To satisfy its store product needs, it requires quick response from its vendors. KG typically orders smaller amounts than it did in the pre-EDI environment. Likewise, it orders more often. Today, it orders every 2 weeks to refill stock levels and expects shipments quickly enough to never run short of any items.

In order to facilitate receipt of product at the stores, two of KG's major vendors, Levi Strauss & Company and Haggar Apparel Company, generate and transmit an ASN to KG as soon as the KG order has been picked and is ready to ship. The ASN contains all the information KG needs to receive the product, such as delivery schedule, identifying code of each shipping carton, and the items contained in each. Upon receipt of the ASN at the central location, KG notifies store managers of when their orders will arrive and which items are contained in the shipment. The store manager then schedules staff for receiving the delivery and plans store space for immediate display of the merchandise.

To further facilitate the process, items received from Levi and Haggar are pretagged with a bar-coded label on each item that contains KG's item number and retail price. During the receiving process, the bar codes on the shipping cartons are matched to the identifying codes in the ASN to verify that the correct merchandise has been received. Individual items are carried to the floor and are hung directly on the racks. As they are unpacked, quantities received are counted and entered or scanned automatically and placed into the store inventory file.

KG has commitments from all vendors to support UPC premarking in the future. KG plans on verifying the ASN against receiving information to automatically generate payments in the future. It already has an automated system of matching and reconciling paper invoices. Even though it still has to key-enter invoice information prior to reconciliation, it does not intend to further automate this process through the use of EDI invoices. Instead it plans on using the ASN as the trigger for payments. So far, it has no plans to support EFT or the EDI payment remittance transaction. Because bottom-line profits were KG's main concern, it put its initial efforts into developing business systems from which it anticipated the greatest savings.

KG continues to support the X12 standard and credits VICS, who focuses on the needs of the apparel industry, on accelerating the EDI process for the industry.

Haggar Apparel Company

Haggar is the leading manufacturer of men's dress slacks, sport coats, and suit separates. It also has an extensive line of men's and boy's cotton clothing. It views EDI as a tool to implement its QR program.

Because most of its manufacture is done domestically, its costs of manufacture are higher than those of its offshore competitors. It is looking for ways to level the playing field given the cost differentials that it experiences. Its strategy is to provide excellent service and quick replenishment of stock to attract retailers to its product line.

About 6 1/2 years ago, Haggar was first asked to accept a customer's purchase orders in its own proprietary standard. It agreed but had to develop a front end for its order-entry system to accept and process them. Today, it strongly favors supporting the industry-accepted ANSI X12 standard. As an active member of VICS, it provides input on behalf of the general merchandise sector of the retail industry.

Prior to EDI, the Haggar sales representative regularly visited his or her accounts, took a physical inventory, and generated an order based on the number of items remaining on the racks mitigated by such variables as seasonal adjustments. Merchandise was sold to the retailer by lot; so for each style ordered, the retailer automatically received a predefined assortment of sizes and colors. The buyer at the retail location would approve the order and the sales representative would return to Haggar with it. Because sales representatives visited multiple customers during one sales trip, it could take from 7 to 10 days before the order was received and key-entered into the Haggar order-entry system.

Typically, this process would be carried out monthly, which meant that the retailer was required to carry a full month's inventory of items. If a particular size or color sold out, the retailer was stuck until the next order arrived. Since most retailers kept no detailed sales information on the size and color level, they were not in a position to make smart buying decisions.

More recently, retailers capture POS information at the size and color level for each stock-keeping unit (SKU) and order to more closely meet customer demand. This puts some pressure on the manufacturer to have the required colors and sizes of each style on hand to meet retailer demand.

In order to support its retail customers Haggar implemented a QR program. As part of this program, it regularly sends and receives several EDI transaction sets, including the forecast, purchase order, purchase order change, invoice, ASN, product activity data, POS data, and inventory advice. In addition, Haggar began to premark items by tagging them with the UPC number. The retailer can use ticket information as the entry to its POS system, either scanning the bar-coded representation of the code or key-entering the code itself. Also as part of QR, Haggar implemented a replenishment system to assist retailers to order quantities to more closely match their demands. When the retailer furnishes either POS or inventory-on-hand data to Haggar, the replenishment system develops a retailer order to replenish stock up to the model stock configuration agreed to by the retailer. The system can handle not only an ongoing or normal stock configuration but also stock levels for special promotions for up to three dates.

Success is easily measured; the system tracks initial stock configuration by style or model number as well as ongoing sales of Haggar products. Significant results have been noted; there has been an increase in sales of Haggar items with an accompanying decrease in stocking levels. Haggar today handles about 80 percent of its shipments using EDI transmissions, some through receipt of an EDI purchase order and others through receipt of a POS transaction set and internal generation of a suggested order. Sales representatives still travel to non-EDI customers to generate orders.

Haggar credits VICS with a great deal of the success of EDI in the apparel industry. This organization has made EDI visible, pointing out the benefits to participating companies. It has also acted as a buffer between the ANSI X12 standards organization and standards users in the retail industry. This allows manufacturers and retailers to focus on the data requirements for their own business situations.

In selecting EDI trading partners, Haggar found that those already doing EDI with other suppliers were the easiest to bring up. It found that most of its EDI-active customers, typically the large retail chains, have bought mainframe computer software to translate between their internal or application format and the VICS subset of ANSI X12. Haggar itself has installed mainframe computer translation software as well. Most of its internal effort has been directed at making orders flow faster and more efficiently. It considers EDI a tool to facilitate the flow of information between itself and its trading partners. It expects all savings to be realized through innovative use of timely and accurate EDI data.

Haggar has noticed that the implementation of EDI at its customer sites often takes a back seat to the implementation of POS systems for tracking of sales. However, Haggar believes that there are enough benefits from EDI to justify implementing EDI prior to installing a POS system. It currently has some store chains sending and receiving EDI data with no POS system. While these companies must still do manual inventory counts and are not part of the QR program, they do realize the benefits of eliminating printed hard copy, mailing and handling paper documents, key entry, and internally generated errors.

For those in the QR program, Haggar offers priority handling within its system. The result is that Haggar is able to provide fast turnaround and the customers are able to order smaller quantities at shorter time intervals. Most order weekly or biweekly; some even send sales data daily.

For the majority of its department store customers, Haggar packages and ships directly to the retail store. As part of the shipping process, it scans the contents of each shipping carton and generates a packing slip which it places in the carton. The retailer compares the packing slip to its physical count of items in the carton to verify that it has received the correct ordered quantities.

Today, 98 customers, representing about 10,000 stores, receive ASNs from Haggar. While 51 others are receiving electronic invoices, Haggar reports that the invoice has been slow to catch on. Haggar anticipates that the ASN will continue to grow and will be used by retailers in a dual role: one, to report on an impending shipment and, two, to act as an invoice.

Additionally Haggar transmits price and sales catalog data via EDI to some trading partners. These transmissions contain updated prices of the SKUs that the trading partner regularly orders. As with Levi, it has not initiated any EDI/EFT or payment remittance activity. Likewise, there has been no EDI activity with carriers.

On the vendor side, Haggar receives ASNs from its major fabric suppliers. Potentially, this can save them a good deal of time during the receiving process. Here's why; traditionally, when fabric is received, the width is measured, the fabric is tagged with a bar code, and a corner of the fabric is cut off to be sent to the lab and analyzed for color before the fabric is stored. This process takes several days to accomplish, so inventory that is received is not freed up for use in new product until a week after its receipt. Today, as part of the ASN, the supplier provides length and width and the results of its own color analysis. In addition, the supplier tags the fabric prior to shipment. Upon receipt, the retailer scans the bar-coded tag and immediately stores the fabric and frees it for use in new product. Haggar saves both time and money by getting detailed information on incoming goods while at the same time reducing its inventory requirements. Today it is ordering 90 percent of its fabric via EDI.

Haggar transacts its EDI business through a third party and encourages its trading partners to do the same. It reports that even though the major manufacturers in its industry such as Levi and itself made the initial move toward EDI, today retailers are usually the drivers of EDI. Each side has excellent economic reasons to encourage EDI with its trading partners because EDI provides a win-win opportunity where both sides benefit.

Summary

The retail industry has been one of the fastest growing segments for use of EDI. Not only have more companies joined the already EDI-active user community, but those doing EDI have increased the number of different EDI transactions that they support and the number of partners and percent of business that they transact electronically. While the grocery chains got an early start in the EDI arena, mass merchandisers have passed them by with greater breadth and depth of EDI usage within their organizations. Likewise, QR programs have been implemented more frequently than have ECR programs. This is due in part to some noticeable differences between the grocery retailers and their general merchandise counterparts. For one thing, mass merchandisers deal with a much larger number of vendors. The Salmon study relates on average 4500 vendors of mass merchandisers to the grocery chain's 1160 vendors. For another, the study relates that mass merchandisers carry a much larger number of SKUs, about 400,000 to 500,000 SKUs to the grocer's 28,000 to 30,000. This makes it harder for mass merchandisers to control their inventories. Finally, the rate of replenishment is vastly different between the two. Here the grocery rate is much higher, turning its stock at five units per week to the mass merchandisers one unit every 3 or 4 weeks. At such a high turnover rate, grocery stores almost always count on their own warehouses or distribution centers for day-to-day replenishment of goods.

Another major difference in the two industries has been their perception of the use of electronically available data, both intercompany and intracompany. In the grocery industry, collection of POS data, scanning, and EDI have been looked at as cost reduction and productivity improvement strategies. In the general merchandise segment of the industry, investments in technology have been considered the conduit for the rapid and reliable flow of information and product. By using information creatively, mass merchandisers have positioned themselves to better respond to consumer needs and capture market share.

Aside from growth in technology, both grocery chains and general merchandisers realize that they will only get a real competitive edge by being the most responsive and offering the best service to the consumer. By building a partnership between retailer and manufacturer, both organizations stand to realize the largest overall benefits. Hence, QR systems are a combination of technology and business services. They are based on the development of partnership goals and the provision of overall system efficiency throughout the supply chain. Efficient consumer response systems, just now beginning to grow, will have the same components and will do the same.

While QR was considered state of the art just a few years ago, today it is almost a requirement for large retail suppliers. Those that do not enhance their service levels to include QR services will find themselves at a marked competitive disadvantage. In instances where brand recognition is not a major factor in sales, the supplier that provides the higher level of service will grab shelf space from its less innovative competitors. The result will be a loss of market share for less innovative companies as well. The same will likely be true for some grocery products. However, for a large portion of goods in this industry, manufacturers have built brand recognition through heavy advertising campaigns. Even when this is true, there is still a good deal of room for brand switching based on price differential. Manufacturers that are able to cut internal costs and reduce prices to the retailers can count on consumers switching brands in their favor.

9

EDI in the Health Care Industry

Just a few years ago all EDI activity in the health care industry centered around the wholesaler and manufacturer of pharmaceuticals and was almost entirely focused on the transmission of the purchase order and the charge-back transaction sets. Today, there is more EDI activity in the wholesaler-manufacturer arena due to the increase in partners and percent of business for those same applications because of the introduction of new EDI applications. However, the types of health care industry players that have begun to use EDI to streamline business procedures and obtain accurate, machine-readable information has increased dramatically.

Considering the amount of attention and press aimed at the health care industry, it is no surprise that EDI and electronic commerce show signs of rapid growth here. The Health Care Financing Administration (HCFA) has set a goal of having 90 percent of Medicare claims automated by 1995. The Work Group on EDI (WEDI) formed by Health and Human Services Secretary Sullivan along with major commercial insurance payers, identified a $10 billion potential savings from the use of EDI and has challenged health care players to make use of EDI by 1994. The department of defense began accepting bids via EDI in July of 1992. Vendors to the Veterans Administration Hospitals now submit their invoices via EDI to ensure payment; EDI purchase orders will be required within the next year. In addition, major multihospital corporations have EDI programs to purchase supplies and sometimes to administer contracts with major suppliers.

There is also a ground swell of interest in EDI by various health care–related associations. It started with NWDA back in the mid-1970s.

Today there is the Health Care EDI Company (HEDIC), Health Care Finance Management Association (HFMA), and Health Industry Business Communications Council (HIBCC) just to name a few. In addition, the ASC X12 Insurance Subcommittee is the largest and one of the most active work groups. It is dedicated to the development of EDI transaction sets for the insurance industry.

In this chapter we'll look into what wholesalers, manufacturers, retailers, health care providers, health care payers, and health care third parties are doing with EDI and related technologies.

We'll start with a discussion of wholesaler-manufacturer EDI applications as they represent not only the first use of EDI in the health care industry, but one of the first and most successful implementations in any industry. Next, we'll look at the uses of EDI by retail pharmacies and drug chains. They are experiencing the same business issues as grocery and general merchandise retailers and therefore are concentrating on similar solutions to facilitate inventory control and replenishment. In fact, all these retailers are also implementing EDI in their pharmacies, where they use it to verify patient medical coverage eligibility for prescription drugs. We'll take a look at hospitals, briefly identifying their uses of EDI and then focusing on a hospital chain that has reaped savings throughout its enterprise's innovative uses of bar-coding. Finally, we'll discuss EDI for health care payers such as insurance companies, Medicare, and Medicaid, describing the various EDI transaction sets that are currently available for use and the activity that we have seen thus far.

Wholesaler-Manufacturer EDI Applications

The relationship between the drug wholesaler and manufacturer is strong, as the wholesaler serves as the distributor of product to hospitals, independent drug stores, and chains. Even when prices are negotiated directly between hospitals or buying groups and manufacturers, the wholesaler is still responsible for distribution. Figure 9-1 illustrates the trading partner relationship between wholesalers, manufacturers, and hospitals.

Historically, as far back as 1972, Bergen Brunswig, one of the largest drug wholesalers, began using the remote computing services of Management Horizons Data Systems (MHDS) to develop purchase orders in print format. Bergen would then print and mail them to the manufacturers. In 1972 a pilot was begun with Eli Lilly in which the purchase

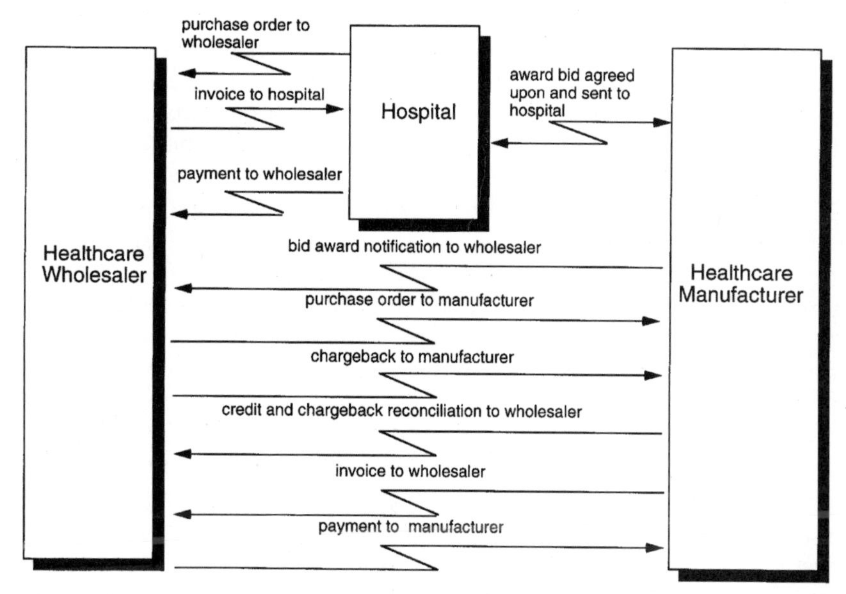

Figure 9-1. Trading partner relationship between health care wholesalers, manufacturers, and hospitals.

orders, in data format, were transmitted directly from the MHDS data center to Lilly. In line with the technology of the day, the data format developed was comprised of fixed-length data fields and 80-character records and contained all the data necessary to conduct business.

At about the same time, the Business Systems Committee of the National Wholesale Druggists Association (NWDA) met to discuss this same possibility: ways of transmitting ethical drug orders directly to manufacturers. As this organization represented wholesale druggists, it was under some pressure to help its membership alleviate pressing business problems. The wholesaler was searching for expense-cutting measures to reduce internal costs. This, the wholesaler believed would counteract its unacceptably low return on net worth (RONW) of under 10 percent and net profit after taxes of under 1 percent.

Implementing this new technology to electronically transmit purchase orders promised to reduce order lead time, which in turn would decrease inventory and safety-stock requirements. In fact, implementing EDI was one of the prime factors in growing RONW to over 15 percent through 1984 and net profit margin to 1.25 percent over the same period. Over the next few years, those figures dropped because operating expenses decreased at a slower rate than gross margin. The year 1987 showed a turnaround of those figures again.

By 1974, MHDS had developed the system to which it attracted 16 drug wholesalers in 32 locations. The system transmitted suggested order quantity information to drug wholesalers for review by their buyers. The first manufacturer to pass information through this system was Eli Lilly, followed by Mead Johnson, Burroughs Welcome, and others over the next 2 years.

In 1975, Informatics General Corporation, which had purchased MHDS, approached NWDA with the idea of an industry clearinghouse for electronic purchase orders. The NWDA Business Systems Committee was enthusiastic. After receiving responses to a request for proposal sent to several potential third-party providers, the committee selected Informatics as the EDI third party to run its pilot program. During the initial program, Informatics and NWDA worked together to define the information needs of the industry and to develop the EDI standard. They selected the purchase order and invoice transaction sets for conversion to the electronic form. The standard that they developed had fixed-length fields and fixed-length records. It came to be known as the ORDERNET/NWDA standard.

In 1978, NWDA endorsed the Informatics ORDERNET clearinghouse system, which prompted Informatics General to name its new division the ORDERNET Services Division. All manufacturers from the initial program became clearinghouse customers, except Eli Lilly which continued to receive orders directly. Those original manufacturers, most other pharmaceutical manufacturers, and almost 100 percent of the wholesalers in the industry are today active EDI users. Most still use the same EDI network. Informatics was purchased in the early 1980s by Sterling Software and that network is now called COMMERCE:Network. Even Eli Lilly, who continued to support direct EDI transmissions for many years, has recently started transmitting EDI transactions through the network as well.

With the endorsement of the NWDA the following occurred:

- ORDERNET Services agreed to enhance the ORDERNET purchase order formats.
- ORDERNET took on the responsibility of sales and marketing of the system.
- ORDERNET started selling EDI services to industry buyers and sellers, with the strong active support of the NWDA. Richard C. Cook, then director of operations and research for NWDA, was a strong believer in the system and strongly supported the program.
- NWDA gave ORDERNET the opportunity to present the program to wholesalers and manufacturers at trade shows and meetings.

- ORDERNET agreed to refrain from marketing to any entity competitive to wholesalers, namely hospitals and chain drug stores, for a period of 2 years, which gave the wholesalers a differential advantage in the marketplace.

On the strength of this existing system, the American Surgical Trade Association (ASTA), whose membership is made up of distributors of medical and surgical products, approached ORDERNET Services about a year later to develop a similar system. Not only was it interested in transmitting purchase orders electronically but also in transactions for sales tracing and rebate reporting. A similar fixed-length format, called COMM-NET, was developed for this industry.

It is interesting to note that even with a large degree of commonality between the drug and pharmaceutical companies and the medical and surgical companies, the sales tracing and rebate reporting transactions have never been supported by the drug sector. That is because a similar service, provided by Drug Distribution Data (a division of IMS/America), was already available. Electronic data interchange has been evolutionary, not revolutionary, in this industry, based solely on business need.

There are three main areas of EDI usage in this segment of the health care industry:

- Purchase order transactions sent from the wholesalers to the manufacturers.
- Charge-back family of transactions transmitted between wholesalers and manufacturers.
- Invoices sent from the manufacturers to the wholesalers.

By 1986, 99 percent of drug wholesalers were using EDI and specifically the ORDERNET/NWDA industry-specific purchase order transactions. Ninety percent of pharmaceutical manufacturers were doing the same, representing the largest penetration of EDI usage in an industry.

Electronic Purchase Orders

Let's examine the benefits each of the trading partners receives from sending and receiving the purchase order electronically. The wholesalers are able to reduce order lead time, thereby reducing safety stock and inventory requirements. In addition, they have eliminated time-consuming manual tasks related to handling of paper transactions. By sending complete and correct electronic transactions to the manufacturer, they improve their chances of receiving timely deliveries of the required items.

The manufacturers have also eliminated manual procedures, including interpretation and completion of incoming purchase order information. In so doing, they have eliminated transcription and interpretation errors which, in turn, have all but eliminated shipments of incorrect product and returns based on incorrect shipment. In addition, they are able to rapidly process the incoming electronic orders which allows them to send the product sooner than was possible prior to EDI. This, in turn, improves the customer service that they can provide. Finally, by receiving machine-readable, accurate, and complete purchase order data, they have set the stage to disseminate this information to various computer applications throughout the company, such as the billing application for the generation of an EDI invoice, and to strategic personnel, such as auditors, functional business managers, and top-level strategy planners to improve business procedures and decisions.

Charge-Back Family of Transactions

The development of three electronic transaction sets in the industry-specific ORDERNET/NWDA standard [the bid award notification, the charge-back debit memo, and the charge-back reconciliation (needed to process charge-back claims from the wholesaler to the manufacturer)], represents a departure from the traditional view of EDI. Traditionally EDI has been thought of as a substitute for a standard paper business document. In this case, no paper documents existed. The charge-back process represents a specific business need of the health care industry. It exists because large hospitals often negotiate prices directly with a manufacturer but order and receive delivery through a wholesaler. To reimburse the wholesaler for the price differential between its normal selling price and the lower selling price awarded to the hospital, the manufacturer agrees to credit the wholesaler with the difference.

Pre-EDI, issuing charge-back claims, receiving credits, and reconciling charge-back activity had been a long and arduous procedure. With EDI and automated systems, this procedure has been substantially streamlined and shortened.

The first of the three electronic charge-back transactions to be used in large volumes is the charge-back debit memo. Generated by the wholesaler, it contains details of the hospital's purchases needed to initiate a return credit from the manufacturer. The wholesaler uses these credits toward future purchases from the same manufacturer. Almost all charge backs today are transacted electronically.

Unfortunately, in most cases, the wholesaler learns that the hospital has negotiated a special price with a manufacturer after its first invoice has been rejected. In order to eliminate this problem, the wholesaler requests that the manufacturer send it an electronic bid award notification transaction as soon as the manufacturer awards a special contract to the hospital. In that way, the wholesaler can be aware of special pricing ahead of time and can invoice correctly the first time. This, in turn, improves its chances of being paid in a timely manner. Manufacturers are beginning to develop applications to generate the bid award transaction, both to comply with the wholesaler's request and to eliminate the manual procedures they experience when logging contracts and tracking charge-back debit claims and to reduce their staffing requirements as well.

Finally, there is a third transaction to complete the set, the charge-back reconciliation, that debits or credits the wholesaler based on the original charge-back debit memo and the hospital contract terms. This transaction is also gaining in popularity at this time.

The Electronic Invoice

The third area of EDI activity in this segment of the health care industry is the invoice transaction set. By retaining a copy of the electronic purchase order, the wholesaler has all the information it needs against which to reconcile incoming invoices. If these invoices are received in paper form, the information must be interpreted and key-entered, a process which is fraught with errors. However, with electronic transmission of machine-readable data, an invoice reconciliation application can be developed which would allow the wholesaler to reconcile manufacturer invoices at the line item level. By automating the reconciliation procedure, the wholesaler has a good deal to gain: it would be in a position to authorize payment sooner, take advantage of discount terms more often, and decrease its cost for product. Since wholesalers experience such low net-profit margins in this industry, there is a great deal of appeal to lowering costs in this way. In addition, the wholesaler could greatly reduce its accounts payable staff. Unfortunately, it is manufacturers who generate the invoice, and because they are typically high-profit companies, realizing about 25 percent net profit, they rarely initiate generation of this transaction until pressure has been applied by the wholesaler. Therefore, usage of the electronic invoice has not, at yet, reached anywhere near the volume of either the purchase order or the charge-back transaction.

Following is the flow of information in the electronic environment for the invoice transaction:

- The invoice is generated out of the vendor's computer system using both purchase order and shipment information.

- The invoice is transmitted electronically to the wholesaler in standard data format.
- VAN control reports verify that the invoice was sent through the network and was retrieved successfully by the trading partner.
- Machine-readable invoices are automatically reconciled against the wholesaler's open purchase order and receiving information to determine whether they are correct.
- Payment is automatically authorized, and either a check is cut or the payment is transferred electronically in time to take advantage of discount terms.
- Incorrect invoices are handled by accounts payable staff as exceptions.

Of course, in order to effect this flow with automated processing the wholesaler must have developed the necessary applications to:

- Preserve a file of open purchase orders to which accounts payable has access.
- Update inventory with receiving information that can also be accessed by accounts payable.
- Reconcile the invoice at the line item level against both the purchase order and the receiving information.

There is some additional usage of EDI in this industry segment. Some manufacturers are exchanging electronic shipping and billing transactions with carriers in order to obtain freight charges in a timely manner, to include them on the original invoice to the wholesaler, and/or to speed up authorization and payment of the bills.

In addition, some innovative wholesalers have initiated payments in the form of EFT. Many companies are leery of using EFT because they believe it will lead to loss of float. Technically speaking, this is true. However, loss of float does not necessarily mean losing use of the money any sooner. When using EFT, the payer can specify the payment date. That date can be the same as the one on which the funds would have cleared using traditional paper checks.

Electronic data interchange was initiated in this segment of the industry because of economic hardship of drug wholesalers. The extremely low profit margins experienced by drug wholesalers prompted them to generate and transmit the purchase order and charge-back debit memo, both of which have become the de facto mode of doing business. However, when it came to the transactions returned by the manufacturer, adoption has been much slower. While almost all wholesalers have been transmitting orders and charge-backs for several years, as of a year ago, little more that 30 manufacturers were prepared to return the invoice and the related charge-back transactions. There are still substantial benefits to be derived from EDI in this industry. NWDA has continued to be an active participant, encouraging its wholesaler members and manufacturer associate members to increase their participation in EDI by transacting business electronically in any and all business areas in which they can realize the benefits of decreased costs, staff reductions, and savings of time.

In this industry, just as in retail, cooperation between buyers and sellers has grown and has become the basis of programs aimed at optimizing business of both parties. An example is Kimberly Clark, who offers vendor-managed inventory programs for its major distributor customers such as Owens & Minor. These programs are geared to optimizing the amount of product in the supply chain in order to minimize total time and costs in the replenishment cycle. With a program such as this, the distributor no longer needs to generate and send purchase orders. In addition, it no longer needs to reconcile purchase orders, receiving information, and invoices. Through a series of EDI transaction sets, O&M relates their current inventory position and sales data. Kimberly Clark analyzes the data and develops a suggested shipping list. Kimberly staff may modify the suggested list to minimize shipping costs. Upon release of the shipment, O&M is sent a purchase order acknowledgment to apprise it of the impending shipment and an invoice. O&M uses these incoming EDI transaction sets to update its inventory and accounting systems and to release payment.

With the breadth of EDI transactions available and the widespread acceptance of the ANSI X12 EDI standard, the NWDA strongly encourages its members and associate members to adopt X12 and phase out ORDERNET/NWDA applications. The trend has been for new applications to be implemented using the X12 public standard. One factor that has acted to encourage this switch to X12 is the large number of different transaction sets available in the X12 standard. A vendor-managed inventory program, for example, would be impossible using only the ORDERNET/NWDA standard because of its limited number of electronic transaction sets. On the other hand, there are factors that are causing companies to retain the ORDERNET/NWDA standard for existing EDI applications. The main one is that companies supporting the ORDERNET/NWDA standard have already programmed their application link and translator and, because of the simplicity of the standard, have no need to purchase an EDI translator.

EDI for Independent Drug Stores and Large Drug Store Chains

Chain drug stores are experiencing all the same business issues as other large retailers. This being the case, they are responding with similar EDI solutions. Typically, the chain drug store uses the X12 standard to transmit purchase orders and receive invoices. Some are also receiving ASNs. To the extent that chain drug stores also carry grocery and general merchandise items, they are supporting UCS for the former and X12 for the latter.

Aside from these typical transaction sets, another EDI application has grown rapidly in the pharmacies of independent drug stores and drug chains. Today, over 95 percent of all pharmacies are computerized. Many already have sophisticated second- and third-generation systems. These systems assist the pharmacy to order, some using the American Society for Automation in Pharmacy (ASAP) guidelines and the EDI standard; others using various proprietary standards. A second pharmacy application facilitates the handling of prescriptions by acquiring the customer's eligibility and coverage information prior to filling the prescription. There were over 1.7 billion prescriptions filled in 1993 alone. Half of these are reimbursed by a third-party payer. More than 50 percent were submitted interactively (real time) for payment. When the pharmacy receives the prescription, either via physician call or patient carry-in, the required information is entered into the on-line system using the National Counter Prescription Drug Plan's (NCPDP) pharmacy standards for real-time transactions. During this session, the pharmacy identifies the patient and describes the prescription drug, dosage, and quantity. This gives the pharmacy access to a database of insurance coverage from which it is possible to determine whether the patient is covered for this drug and how much the payer will pay. If there is a copay, the system reports back what the copay amount is. If generic substitutions are allowed or required, the pharmacist learns this from the system as well. The ASC X12 Insurance Subcommittee Working Group SPWG3 has developed two EDI transaction sets to handle this application. The 257 handles eligibility and benefit inquiries. The 258 handles the response. These transactions are also interesting because they are the first to combine an X12 transaction format with the interactive UN/EDIFACT control envelope format. The transaction and envelopes combination was recently accepted by a public vote in the spring of 1994. Pharmacy systems are helping to improve buying and speed up the processing of prescriptions.

EDI and Bar-Coding for Hospitals

While EDI is used in hospitals, its growth has been less than spectacular. It has grown in isolated pockets. One reason for this is the distributed processing environment of most hospitals. While corporations have moved toward centralizing of their purchasing function, hospitals have retained separate systems and authority for the purchase of supplies. For example, hospital pharmacies are separate from cafeterias and from nonethical supplies as well. Another is the ready availability of vendor on-line systems. Most large vendors of hospital supplies have a proprietary, PC-based system that they are happy to place at the hospital location to be used for ordering their own products. Over time, the hospital has accumulated several of these systems and has incorporated their use into their daily ordering routines. Even though many proprietary ordering systems are certainly not the ideal purchasing solution, they continue to be used because they are there and already in use. In fact, handling the procurement of product via EDI is fertile ground for savings by hospitals. Recent articles report that the error rates being experienced by hospitals in the procurement process can be as high as 30 percent. With the cost of a hospital purchase order ranging between $35 and $85, from $10 to $25 can be attributed to errors alone.

There are other innovative uses of EDI and related electronic commerce technologies being used in hospitals today. A recent article in *EDI World* describes the many uses that the 313-bed St. Alexius Medical Center in central North Dakota is making of bar-coding.

Back in 1987, St. Alexius began using bar codes to track flow of product. It applied piggyback bar-code labels to all products that flowed through its distribution centers. It established par levels for each product so as to control inventory in 49 stock locations throughout the facility. When a product is used, its bar-coded label is transferred to the patient's charge card. These labels are then scanned into computers which update the patient's bill as well as keeping track of the product restocking needs. After installing this first bar-code application, which had cost $50,000 for the system, St. Alexius began to track savings. It was discovered that they experienced a savings of $50,000 within the first 6 months of use. As a side benefit, distribution- and patient billing–related errors were all but eliminated. In addition, inventory was reduced by an average of 15 percent in the patient care supply areas. Because of the user-friendliness of the system, one full-time position was eliminated and responsibility for supply distribution was transferred to aides. In all, there was a cost reduction of $24,000 annually. Since the introduction of this system, St. Alexius has increased its use to other supply locations. Today, its system is active in 80 locations throughout the hospital.

Once St. Alexius installed bar-code technology, it began to look for additional applications for its use. Bar-code software is used today in its printing department where preprinted numbers on forms are converted to bar codes and are scanned into files instead of being key-entered. Key-punch staff was provided with scanners. It is estimated that the staff has eliminated 7,000,000 key strokes and has almost 100 percent accuracy.

In addition, their laboratory department uses bar codes extensively. The laboratory system generates bar-coded container labels to correspond with each lab test requested. For example, as blood is collected, the label is attached to the tube. This instantly associates the patient with the blood sample. When the bar code is scanned, the same analyzer that does the scan automatically performs the correct test and reports the results back to the lab information system. This system in turn interfaces with accounting and updates the patient's bill. Results are also electronically transmitted to the ordering physician's office.

In the microbiology department, bar codes have eliminated 85 percent of the key strokes needed to enter test results for blood, urine, and throat cultures. This has reduced the time required by 80 percent as well.

In the hospital pharmacy, the results have been exceptional. St. Alexius has installed a system to prepackage dosage quantities of its 212 fastest moving tablet and capsule drugs from the bulk containers received from the manufacturer or wholesaler. A bar-coded label is applied to each dose package. Between its bar-coded prepackaged dose packages and those prepackaged and bar-coded by manufacturers, St. Alexius is able to dispense almost 100 percent of oral tablets and capsules in bar-coded dose packages.

Some of its patient rooms are equipped with radio-frequency bedside computers. When supplies or medications are used, the nurse scans its bar code into the computer. The system immediately identifies the medication. This greatly reduces the chance of administering the incorrect drug. It then upgrades the patient's record and hospital inventory files and charges the drug to the patient. In these rooms, the piggyback bar-coded dose labels are no longer needed, so long as the medication is prepackaged in bar-coded dose units.

With the accuracy of this system, they are even able to confidently match up a dialyzer with a patient and to reuse the dialyzer cell up to 50 times with no chance of error. The ability to reuse expensive dialyzer cells saves them a substantial amount of money.

In the receiving area, St. Alexius has begun to scan incoming cartons that have been bar-coded by the manufacturer. This has increased the accuracy of the receiving process by correctly recognizing incoming products as well as by automatically upgrading purchase order and inventory figures and passing accurate receiving information to the accounts payable system. In addition, St. Alexius has realized a decrease in the amount of time it takes for scanned product to actually reach the stocking locations, ready for use.

Currently a system to track time and attendance of staff is being installed. Each employee will receive a bar-coded identification card. The employee will swipe the card when he or she arrives and again when he or she leaves. This will automatically upgrade attendance records and generate the correct payroll check.

Finally, St. Alexius is researching the possibility of bar-coding its medical forms. After being bar-coded, completed forms can be optically read and cataloged automatically by the system. This in turn will greatly reduce filing errors and access time of patient records.

With all these bar-code applications in place, St. Alexius has realized an increase in efficiency, quality, and accuracy. It continues to search for additional uses of this technology to further streamline its procedures.

EDI for Health Care Payers

Electronic data interchange has been designed primarily for batch transactions, where many transaction sets are grouped together for transmission and are picked up and processed at a later time. However, health care payers are being asked to support interactive EDI applications, where the user is logged on to a computer system and receives immediate and sometimes interactively conversational responses to queries.

Health care providers are asking for information regarding the patient's clinical history as well as payer delivery and procedure coverage. Payers do not object to this timely access of information because they are interested in letting the servicing physician know which health care services will be covered. The Insurance Subcommittee of the ASC X12 committee has been hard at work designing immediate inquiry and response transactions since August of 1992. The transactions are similar to other ANSI X12 transaction sets except for a few differences. For one, the subcommittee has chosen to wrap the interactive transactions in variable-length EDIFACT headers and trailers instead of the longer fixed-length X12 envelope. For another, segments have been purposely made very concise, with limitations on repetitions of groups of segments. The number of mandatory segments and data elements has also been kept to a minimum. All this serves to save time and money and make the interactive processing of these transactions more practical.

While these transactions are being designed as a solution to the instant information needs of the health care industry, interactive requirements and immediate response transactions are by no means limited to this industry. Once approved, these transactions will serve as models for other interactive business requirements.

The insurance subcommittee released these transactions into the approval process during 1994. Based on the voting response of the public, the transactions received approval on the first ballot. The subcommittee is especially pleased to have received approval on these transactions so as to derail proprietary solutions that are already being established by individual health care payers. As in the past, using several proprietary standards instead of one public standard would probably be a step in the wrong direction in the long run as it will serve to increase administrative costs rather than decrease them.

Another application on the same front is the overall reduction of administrative costs throughout the health care system. It is estimated that administrative costs consume between 23 and 50 percent of the health care dollar, and it is believed that widespread use of EDI will greatly reduce that figure. Today, only about 15 percent of nongovernment health care claims are processed electronically. However, predictions are for that number to more than quadruple over the next 3 years. In fact, EDI is one of the few nonpartisan and noncontroversial health care issues on the table today. As such, it has gained industrywide acceptance from the private sector and is a component of all the proposed health care reform programs. Estimates for savings on administrative costs, based on the use of EDI, run as high as $4 billion annually. As the patient, we may soon be issued electronic cards which will be used to conduct an on-line inquiry as to our health care coverage when we seek health care related treatment. Physicians will file their claims electronically, and payers will settle on those claims electronically as well.

Summary

Potential savings in the health care industry are tremendous. However, there are some issues that are causing the growth of EDI to be slower than we would prefer. One is that in some cases, the parties do not have the typical customer-vendor relationship resulting in a lack of partnership feeling and cooperation. For example, not only do payers and health care providers not consider themselves partners, they often distrust one another and rarely cooperate to improve interenterprise procedures. In fact, payers in the health care industry have been traditionally so competitive with one another that they have only in the last few years started cooperating in the design of EDI standards. While in the past, health care payers supplied providers with proprietary systems, competition continues today in the very different ways that they choose to use the standard.

Two, it is the non–computer-oriented health care provider who needs access and input to the various health care payer systems. And, while health care providers are among the most highly educated groups in the country, they are typically not interested in implementing change in their procedures and are less interested when that change requires the use of computers and new computer applications. Electronic data interchange systems for health care providers will only be successful and replace today's highly manual procedures if their high-tech capabilities are completely transparent to the user. This means that the EDI standard, for example, must be developed in the background of a system that requires only commonly used medical codes with which the provider is already familiar. While this approach is possible, it runs counter to the method most often used by software vendors in corporate America.

Another factor in the acceptance of electronic handling of health care information regards the confidentiality issue and the legalities surrounding the retention of paper and the authorizing signature.

Balancing out all these issues is the spotlight on health care reform and the need to reduce administrative and other health care–related costs. The pressure is on for change. Not only will EDI be used more and more throughout the industry, but other electronic commerce technologies will be as well.

10
Financial, U.S. Government, and International EDI

This chapter will deal with three applications of EDI. One that uses EDI for financial transactions through the payment process, another that uses EDI to handle transactions between various agencies of the US government and private sector corporations, and still another that uses EDI for corporate-to-corporate international trade.

Financial EDI

Financial EDI refers to the movement of payments and payment-related information via EDI. Companies doing financial EDI are corporations, both buyers and sellers, and their banks. Rarely is financial EDI the first application that they implement. Instead, it follows the conversion to electronic format of nonfinancial transactions. By introducing financial EDI, an organization has the opportunity to eliminate paper from the entire purchase-through-payment loop.

Just as in nonfinancial EDI, the act of transmitting and receiving data electronically does not automatically provide much in the way of benefits. Even automating existing procedures, when they were based on outmoded business practices and technologies, provides only limited benefits. It is only when intracompany information flows and business procedures are reengineered to support business objectives that savings are realized.

Potential savings from introducing EDI into the accounts payable area are:

- Reduction of paper check stock
- Elimination of its storage, printing, and postage requirements
- Automation of the reconciliation of incoming invoices to open purchase orders and receiving information

In addition, companies that implement financial EDI into their accounts payable area improve control of their investment income and can more accurately forecast their disbursements.

Similarly, the benefits to be gained from integrating financial EDI into the accounts receivable area are reduction of manual keying of payment information, elimination of internally generated errors, and an increase in the hit rate for correctly applying cash to customers' accounts. Companies also realize improved forecasting for investment income.

In fact, two of the greatest time, people, and dollar hogs existing in the business environment are the handling of invoices and payments. For this reason, the invoice has gotten more than its share of attention. Today, many companies are looking for alternatives to the invoice. Some generate payments from the ASN (that conveys delivery information) and their own product receipt information. One name for such an arrangement is *pay from receipt* (PFR). Major ingredients of PFR are EDI for sharing payment information and EFT for moving the payment from buyer to seller accounts. Especially when the invoice has been eliminated, suppliers require accurate and detailed payment information to allow them to correctly credit the customer's account. The EDI/EFT standards have been designed to hold detailed payment information.

There are a few major differences between implementing nonfinancial and financial EDI. One is the inclusion of the trading partners' banks as parties to the transactions. Instead of just the two trading partners and up to two EDI VANs needed to perform nonfinancial EDI, there are also up to two financial institutions (banks) and the National Automated Clearing House Association (NACHA) regional banks and clearinghouses needed to conduct business via financial EDI. Another difference regards the EDI standard. While nonfinancial EDI can be transacted completely using only one standard, e.g., ANSI X12, financial EDI requires two, X12 for the EDI transaction sets and the ACH standard to move money transfers through the banking system.

Roles of the Bank in EDI/EFT

Let's take a look at the roles of banks in the EDI/EFT scenario and the services that they provide. Some banks are positioning themselves to be particularly valuable to their EDI customers by setting themselves up as *value-added banks* (VABs). In this role the bank acts as a network and is a communications intermediary, just as EDI VANs are. Potentially they can handle both financial and nonfinancial EDI transactions. Even when a bank does not intend to act as a communications intermediary and uses an EDI VAN itself through which it transmits and receives data, it still performs five roles. One, the bank is an educator. In this role, it works with both accounts payable and accounts receivable people to develop the cost-benefit rationale for automation of functional area processes. Two, it is a consultant. Here, the bank assists customers to analyze their business and systems environments and develop a cost-effective solution for handling payments.

Three, the bank is a processor of information and business transactions. In fact, banks have the legal franchise to act as agents for the transfer of value. With no competitors for handling payments, you would think that their position is very secure. However, banks are experiencing an unusual form of competition: their own customers. Here's why. Some industries pass very high volumes of payment-related data through the banking system. They are seeking to reduce the role of the bank as processor and save on bank charges. What they have begun to do is to net out opposing debts over a predefined period of time outside of the banking system. At the end of the time period, they settle the net debt through the bank. The petroleum industry is well positioned to use this plan. Most companies in this industry are both buyer and supplier to their trading partners. Consequently, trading partners tend to pay and receive payments to and from each other. Instead of passing all transactions through the banking system, this industry nets out opposing payments over, for example, 1 month, and then uses the bank to settle the one net payment at the end of the month. This substantially cuts down on the number of payment transactions handled by the bank and the charges to both payer and payee.

As processors of financial EDI transactions, banks are being asked to provide more information to their buyer and seller customers than was included in their pre-EDI reports. Before EDI, banks traditionally provided only enough information to describe balance and transaction processing. Corporate treasury used this information to monitor its movement of cash. With EDI, the bank often receives and is able to pass along to the corporation's accounts receivable department full remittance information. There the incoming remittance data are reconciled against either invoices or shipping data and are used to apply cash to the customer's account. Figure 10-1 illustrates this "data with dollars" approach. Sometimes the payer chooses instead to pass remittance data to its trading partner either directly or through an EDI VAN and an ACH credit, requesting the transfer of funds to the seller's account,

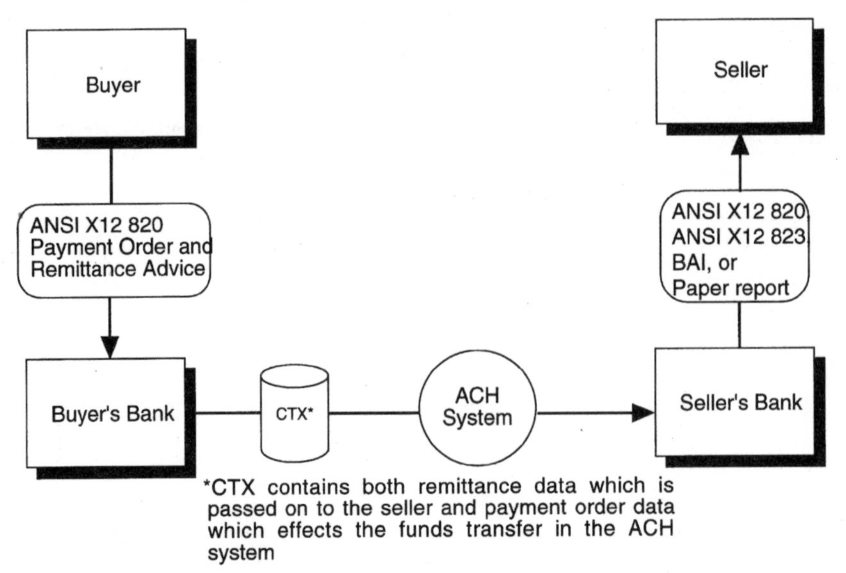

Figure 10-1. Data with dollars information flow.

through the bank. When this approach is used, the seller's bank passes to the seller the payment reference number assigned by the buyer and includes the same number in the ACH transaction. The seller uses this reference number to reconcile incoming payments by reassociating account balances received from the bank with remittance data received from its trading partner. Figure 10-2 illustrates this "data separate from dollars" scenario.

In addition to handling ACH credits which authorize the bank to credit the seller's account, banks also process ACH debits. When the debit scenario is used, the buyer sends an authorization containing an authorization number and payment information to the disbursing (paying) bank via EDI payment remittance. The buyer sends the same transaction with full remittance data to its supplier. The supplier then sends an EDI payment remittance to its bank with authorization to debit the buyer's account. The authorization number received from the customer is included. When the supplier's bank receives the authorization, it attempts to match this authorization number to that received from the buyer's bank. Upon a good match, the debit is posted, which transfers funds to the seller's bank. If there is no match, the debit is returned. Figure 10-3 illustrates the "debit" approach.

There are several advantages of the debit approach. For one, it gives buyers more control over their bank funds. For another, only disbursement (paying) banks need to be able to process debit authorizations. The

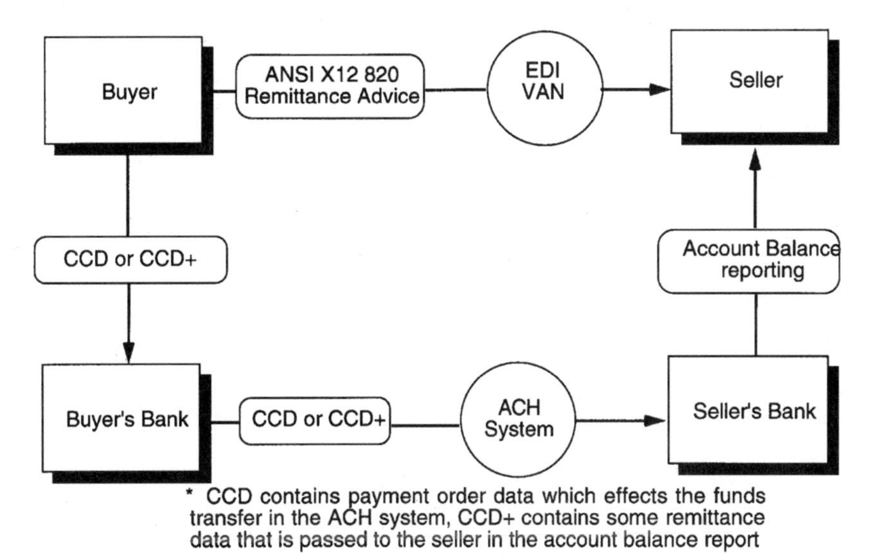

Figure 10-2. Data separate from dollars information flow.

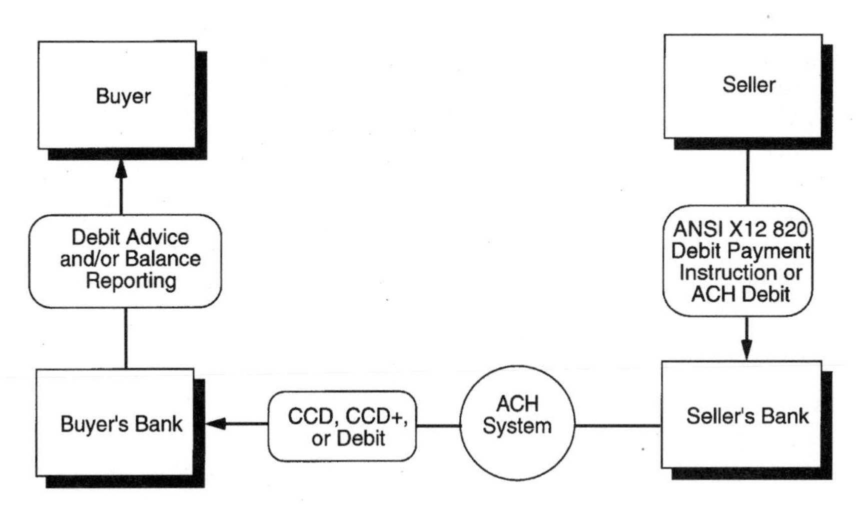

Figure 10-3. Debit approach to payment.

sellers' banks need to process the ACH debit, a capability that all 12,500 ACH banks already have. Unfortunately this is not true for the ACH credit. The vast majority of ACH banks are not capable of handling the data with dollars ACH credit scenario or processing the authorization number used in the data separate from dollars credit. Additional advantages of using the ACH debit to initiate a payment are that it is easier to implement, it requires no payment acknowledgment, it allows improved cash forecasting, it is easier to attract trading partners to this approach, and it eliminates credit problems for the buyer or buyer's bank.

The main disadvantage is that corporate buyers do not typically like the idea of the seller debiting their accounts. They believe that they will lose control over their accounts, even though this is exactly the opposite of what actually happens. With education, buyers should change their bias against the debit.

The fourth role of the bank is as implementation coordinator for their corporate customers. In this role, banks are being asked to assist their customers to integrate EDI into their financial functional areas. Because incorporating EDI data into daily processing tends to be complex, banks need to devote fairly high level resources to this task. Unfortunately, they are finding it difficult to charge for the service as corporations believe that they have already paid for it as part of their general EDI bank fees. The implementation coordinator role is a new one for banks as they have typically not been asked to help in the implementation of treasury management services.

Finally, banks are asked to provide information to the industry at large. In this role they participate in NACHA and local ACH associations. They are also active participants in the X12 process, designing, approving, and maintaining EDI transaction sets. Finally, they speak at local EDI user groups and various trade conferences and write articles for industry publications.

The goal for the EDI-capable bank is to grow its EDI activity volumes to a point where EDI becomes a money-maker for them. This goal is rarely realized today. For one thing, very few banks handle much EDI activity yet. For another, charges for EDI services are very competitive. In fact, banks typically have been undercharging for them, using profits from other more lucrative services to cover their shortfall. As more and more payments are handled via EDI, banks will need to reevaluate their EDI pricing structure and bring charges into line with their internal costs processing.

Comdisco's EDI/EFT Implementation

A recently published case study of Comdisco, Inc., a multinational high-tech management company, shows how it greatly improved the speed and quality of its accounts receivable department by converting paper- and processing-intensive lockbox deposits to EDI payment remittances and reengineered its internal processes. Comdisco's accounts receivable department is responsible for generating and sending 20,000 invoices and receiving and processing 12,500 checks, and 300 lockbox deposits each month.

For each paper-based lockbox deposit, Comdisco performed seven discrete processes, which required seven dedicated resources. Because of the fact that the payment traveled from person to person, its processing was difficult to track. With paper checks, containing only a limited amount of remittance data, the situation was similar. Generally access to data was limited to the person or, at most, to the department that was handling the transaction, and processing consisted of a number of sequential tasks, just as for the lockbox deposits.

Comdisco concurrently implemented financial EDI, and rethought its cash application systems and procedures. It redesigned its systems, developed new measurement criteria, eliminated antiquated business procedures, and changed job tasks and company organization to support the new processing environment.

Implementation required the active participation of various corporate players. Responsibility for this effort was assigned to three parties, one, a dedicated data-processing person, two, an accounts receivable person, and, three, an EDI-capable bank. The staff was empowered to work with the bank to develop a new processing flow. Over a 6-month period changes were implemented. The result was a reduction in accounts receivable and data-processing staff by four full-time positions and a reduction of processing time by up to 50 percent, from 2 to 4 days to 8 hours. Comdisco is able to handle an annual 15 percent increase in number of payments using the same staff. By quicker and more accurate processing, it has improved both customer relationships and satisfaction.

EDI Transaction Sets Used in Financial EDI

There are several EDI transaction sets that are transmitted between the buyer and its bank via EDI. Figure 10-4 contains the ANSI X12 transaction set numbers and names and illustrates the direction in which each flows.

Payment Order/Remittance Advice (820). This transaction may be used as a remittance advice only and transmitted either directly or through an EDI VAN to the seller (payee), as is the case when data separate from dollars is used. Or, it may be used as both a payment order to a financial institution and a remittance advice. In this case, it is transmitted to a financial institution, either directly or through an EDI VAN. The financial institution may either pass the remittance advice to the seller and effect the transfer of funds by generating and passing one of the ACH formats through the ACH clearinghouse, or it may insert the

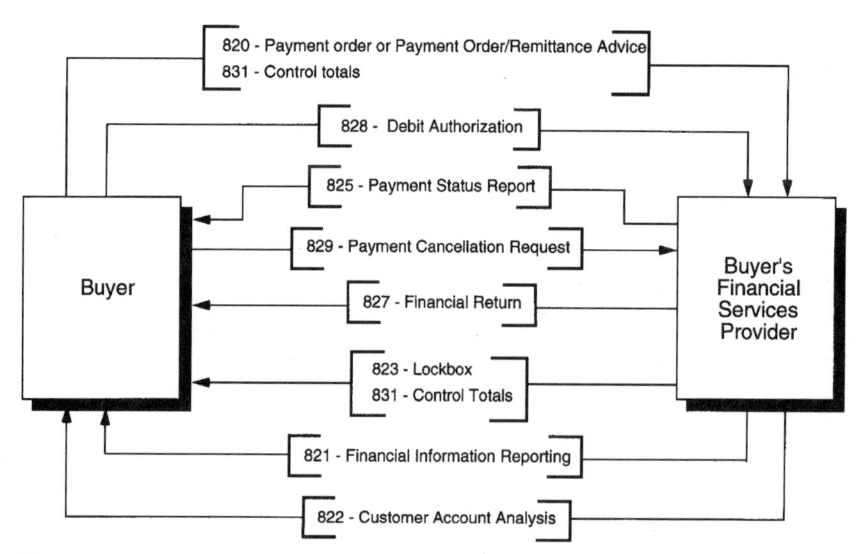

Figure 10-4. ANSI X12 transaction sets between a buyer and its bank and the direction in which they flow.

remittance advice into a CTX transaction and send the entire ACH CTX through the ACH clearinghouse to the seller bank. This is known as data with dollars. The various ACH mechanisms for moving dollars through the banking system will be described later in this chapter.

Debit Authorization (828). This transaction set is transmitted by the buyer (payer) to its bank. By authorizing specific payments for specific dates, the buyer increases the control it has over its account.

Lockbox Transaction (823). This transaction set is used by the bank to report remittance data to buyer customers. Because it can be used for both X12 remittance advice and payment orders as well as for paper checks, it provides a common way for the corporation to receive information via EDI on all its payments.

Payment Cancellation Request (829). This transaction is sent by the buyer to its bank to cancel a previously sent payment instruction, prior to the payment being cleared through the ACH. Because payment instructions can be held until the date specified by the buyer, they can be caught for cancellation if need be.

Financial Information Reporting (821); Payment Status Report (825); and Customer Account Analysis (822). These three transaction sets are sent from the bank to its buyer customer to report on the status of its account or the transactions that it transmitted. The payment status report provides information on the successful completion of a payment. The customer account analysis and financial information report provide overall account information.

Financial Return Notice (827). This transaction set is sent by the bank to the buyer to report on items returned by the ACH system.

Control Totals (831). This transaction set is sent by the buyer to its bank to provide totals for a transmission of one or more remittance advice/payment order transaction sets. It is also sent by the bank to its buyer customer to reflect totals for a transmission of lockbox transactions.

Application Acknowledgment (824) and Functional Acknowledgment (997). These transaction sets flow between buyer and seller trading partners. They may flow through the banking system when one or both banks are functioning as a VAB. They are initiated by the receiver of EDI transaction sets to advise the sender that its transmission has been received and has either been accepted as correct and complete according to the syntax rules of the X12 standard or rejected as incomplete or incorrect. In financial EDI, the buyer acknowledges receipt of the invoice with the functional acknowledgment and the application acknowledgment (if used); the seller acknowledges receipt of remittance advices. The functional acknowledgment (997) is used more regularly than the application acknowledgment (824).

ACH Mechanisms for Transfer of Value

There are several ways to move funds from the buyer's to the seller's account. Figure 10-5 illustrates them. A book transfer is used when both buyer and seller have accounts at the same bank. Using a book transfer, the buyer's account is debited for the amount of the payment. The seller's account is credited for the same amount. When dollars must be credited to an account in another bank, the payment is cleared through the automated clearing house (ACH). Settlement through the ACH requires 1 or 2 days. Over 10,000 US banks and other financial institutions are members of the ACH. There are three mechanisms for making a payment.

1. *Fed Wire.* This is a transfer of funds from one bank to another using the Federal Reserve system. This provides a real-time, same-day transfer of funds. The Fed Wire contains all the information needed

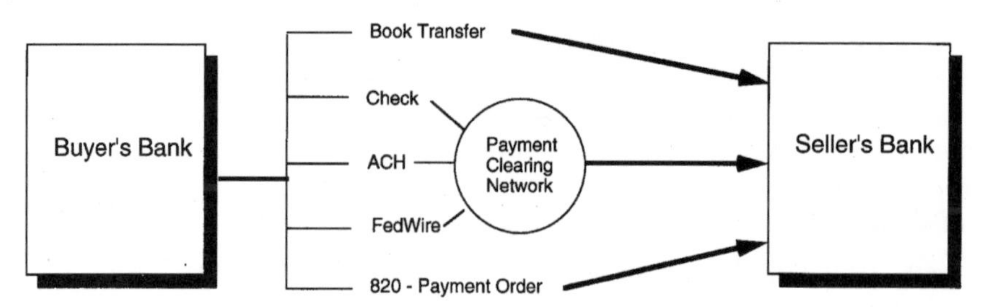

Figure 10-5. Moving funds from the seller's bank to the seller.

to effect the transfer of funds with room for up to 200 characters of free-form remittance data.

2. *Paper check.* This mechanism typically contains even less free-form remittance data in its stub.
3. *X12 820, remittance advice/payment order.*

As mentioned before, when the 820 is used as both remittance advice and payment order, the buyer's bank forwards the remittance portion either directly to the seller or to the seller's bank.

While EDI transactions are used between buyers and sellers and their banks, the NACHA formats are used by the ACH between banks. The ACH formats differ from X12 in that they contain fixed-length fields within fixed-length records. There are several alternative ways of moving a payment through the ACH. The cash concentration and disbursement (CCD) transaction moves the value through the ACH system with only a trace number to allow for tracking. The CCD+ also moves value through the ACH. It contains an addenda record which may hold a limited amount of remittance data, possibly a small portion of a remittance advice (820). The corporate trade payment (CTP) transaction transfers value from buyer to seller accounts and also passes some remittance information in ACH addenda records. This transaction is being phased out as it has never developed a substantial user base. Finally, there is the corporate trade exchange (CTX) transaction, which effects a transfer of funds and contains up to 4990 eighty-character addenda records into which a full X12 820 transaction set may be placed.

As a service, some banks are prepared to generate a CCD on behalf of their corporate buyer customers. When this service is used, the buyer sends the remittance advice (820) only to the seller and the payment order (820) only to its bank, either directly or through its regular EDI VAN. The bank generates a CCD or CCD+ transaction from the payment order data fields and passes it through the ACH to the seller's bank. This allows the buyer to send all its EDI transactions, both financial and nonfinancial, through the same EDI VAN communications interface.

Some EDI VANs are prepared to generate CCD, CCD+, or ACH debit payment instructions from the remittance advice (820) that they receive from their buyer customer. When this approach is used, neither the buyer's bank nor the seller's bank needs to be EDI capable.

As described above, there are three main concerns in handling electronic payments. One, which clearinghouse method will be used, Fed Wire or ACH? The ACH is the more popular because it is far less expensive than Fed Wire.

Two, should remittance data travel with or separately from the transfer of value? When they are transmitted separately, the seller receives advance notice of the impending payment and can resolve exceptions and forecast receivables in a more timely manner. However, when received separately, the seller must reconcile remittance data with account balances through use of the trace number contained in each. When they are transmitted together, no reconciliation is needed. However, buyer and seller banks must support the CTX transaction, the combined ACH and X12 format.

Three, should a credit or debit payment method be used? The credit method advantages are: it is easy to use and does not require what the buyer perceives as giving a blank check to the seller. The advantages of the debit method are: it works just like a check and is less costly for the buyer. Also, there is no need for the seller to send remittance data when it initiates the debit.

EDI/EFT is growing. Today there are some major electronic payers such as the U.S. Government, General Motors, and Sears. Technology is not an issue, and various payment methodologies can be used. The same benefits that companies are realizing for nonfinancial transactions can be theirs for their financial transactions and functional areas as well. However, until a critical mass of EDI/EFT payers is attained, paying electronically will not become the accepted method of transferring funds.

Uses of EDI in the U.S. Government

While there are tremendous benefits attainable from using EDI in various government agencies, historically implementation has been hampered by lack of a strong government directive and absence of government standards. In 1991 this situation began to change with the publication of the Federal Information Processing Standard (FIPS) 161 by the department of commerce's National Institute of Standards and

Technology (NIST). While this regulation does not require the use of EDI, it does stipulate that when EDI is used, it will be in the form of either the ANSI X12 or EDIFACT standard. To back up that directive, the department of defense (DOD) has converted its purchasing and logistics messages into the X12 standard. It has also submitted new transaction sets to the X12 committee for approval as standards. In addition, the Computer-Aided Acquisitions and Logistics Support (CALS) Policy Office's EDI/Electronic Commerce Project Office has worked closely with the DOD and has funded an EDI pilot which provides a single point of contact between the DOD and its suppliers. In fact, the DOD has created a central office to define direction for EDI throughout the agency. While early activity was centered around the purchasing, invoicing, and payments areas, more recently EDI activity has been expanded into the use of bid requests and responses, and ASN transportation transactions.

One particularly interesting application is in purchasing and is being used by the General Services Administration (GSA). There officials are planning the inclusion of new EDI technology in such a way as to make it a viable and cost-effective option for even small companies. The GSA is the main purchasing organization of the government. It recommends but does not require EDI of its suppliers. However, because of its desire and need to run a more efficient and cost-effective purchasing operation, GSA has managed to convince over two-thirds of its top suppliers to accept EDI orders. The vision of a paperless purchasing environment started in President Carter's administration. With the continued focus in the current Clinton administration on cutting costs and using high-tech methodologies throughout the federal government, this vision is all the more pertinent.

With the size and breadth of government activity, you will not be surprised to learn that there are well over 100 EDI projects already under way. In fact, it is estimated that there are millions of EDI transactions transmitted in a year within the government between its various agencies. However, few cross government boundaries and travel between an agency and its private sector trading partners. Likewise, few actually follow Federal Information Processing Standard (FIPS) 161 guidelines or use the X12 or EDIFACT standards; instead they employ older proprietary standards. Because these EDI transactions travel electronically, no matter the standard, they still allow the benefits of speed, improved accuracy, and elimination of manual procedures.

The obstacles to implementation between the government and its vendors are many. For one thing, the government deals with a multitude of vendors, the vast majority of which are not currently tied into the federal EDI network. This means that orders going to these suppliers need to be typed and mailed. For small vendors, implementing EDI requires a low-cost, user-friendly software and communications solution. However, many vendors are decidedly uninterested in making even a modest investment in EDI. Luckily, 80 percent of the GSA's purchases are done with 20 percent of its vendors. The GSA has decided to concentrate on its top 50 vendors, who receive from 1500 to 40,000 purchase orders annually. Another obstacle is the existing legal requirement for contracts to be signed and received in writing. Toward resolving this situation, the general accounting office (GAO) has recently removed some of the restrictions on electronic contracts. Still another obstacle has to do with the requirement that the EDI system must comply with Open Systems Interconnection (OSI) standards. This means that any software used to support EDI by the GSA would not only need to be capable of fulfilling all its EDI requirements but would also need to be OSI compliant. The same is true for the communications solution. In spite of these restrictions, it is extremely important that the government find a standard solution for EDI that can be fulfilled with private-sector software products and communications services, rather than reinventing the wheel with information or communications standards of its own.

Finally, the relationship between the federal government and its suppliers is very different from the relationship between corporations and their vendors. In private business, relationships once established are long-lived and based on continued quality of product or service and price. With large customer organizations making EDI a condition of doing business, vendors are prone to invest in the new technology and support electronic trade. On the other hand, when doing business with the government, vendors must continually compete for new contracts even though they are already providing high-quality products and services and have successfully complied with the terms of their previous contract. Some agencies require that the lowest bidder be awarded a contract or a certain classification of company be awarded a contract, e.g., minority-owned company or small business. The leverage that the government has over its vendor base is limited to the current contract only, and vendors are more likely to be selected on the basis of low price than because of high service level or use of new technology. Consequently the government can impose very little pressure on its vendor base to implement EDI.

One electronic commerce solution being used by GSA is an electronic bulletin board service. This service is an ordering gateway called the Multi-User File For Interagency News (MUFFIN). While originally used to supply general supply schedules, it is currently being used to collect orders from the various government agencies. The GSA then turns these agency requests into orders that it sends out to the appropriate vendors.

Following are three case studies of EDI in various parts of the federal government: one, how the Navy is using EDI; two, how the US Customs Service is using EDI; three, EDI/EFT and the Internal Revenue Service.

EDI and the Navy

The Navy may well be the leader in EDI/EC programs within the federal government. It is finding ways of using commercial, off-the-shelf software products to support its use of new technology and is proactively reengineering its business practices and efficiently managing its programs. The Navy EC/EDI Program Office emphasizes coordination between various departments of the navy to eliminate redundant functionality while encouraging independent program development. It is centrally acquiring the hardware and software needed by the various navy EDI sites to ensure a common, standardized solution for all. This in turn will allow the navy to offer standardized training programs, to share in the introduction of technical innovation, and to implement in the most cost-effective manner.

There are several EDI applications already under way, most related to procurement, contract administration, finance and payment, transportation, supply management, and fuels and base operations. Shipboard EDI is one such application which integrates material and financial management for all types of ships. The navy is installing EDI translators aboard 28 afloat sites and their beach detachments. These will provide a standard system through which requisitions for nonstandard items (those with no national stock number) can be received from ships, maintenance activities, air stations, and other internal navy customers. These EDI work stations will replace the off-line manual processes required to process nonstandard orders today and will also be used to track depot level overhaul or repair requests from afloat units to their storage or overhaul facility of repairable components.

For maintenance, the Naval Supply Systems command has begun to develop an EDI application that uses the 867 transaction set to pass demand data for nonstandard items to ICP technicians.

For procurement, the Navy Fleet Material Support Office is making naval procurement systems EDI-capable. The navy plans on providing regional training for vendors to facilitate all EDI small-purchase procurement initiatives. Similarly, the Aviation Supply Office is also concentrating on the procurement process. It is in the process of branching out to finance, logistics, engineering, and information systems. Today Aviation Supply is electronically transmitting ANSI X12 850s (purchase orders) with 60 major vendors. Just this transaction is saving them 15 days lead time in the ordering process. The remittance advice (820) is also being used, with over 100,000 transactions representing $348 million already transmitted. They are also testing the ASN (856). The navy continues to embrace EDI and EC technologies to improve the quality and reduce the cost of its operations.

EDI and the U.S. Customs Service

The U.S. Customs Service has realized substantial productivity gains through the use of EDI. It has implemented an Automated Commercial System (ACS) which is a computerized data-processing and telecommunications system that links customs houses, members of the import trade community, and government agencies with the customs computer. Industry users file the data required by customs for the importation of merchandise. In return they receive information on the status of their cargo. In addition, industry users pay duties, taxes, and fees electronically, through a treasury-approved clearinghouse bank. Not only does the customs service use this information itself, but it exchanges it with governmental and private-sector systems such as the Bureau of the Census, the U.S. Department of Agriculture, Fish & Wildlife Service, Federal Communications Commission, and Statistics Canada. While participation in the ACS is voluntary, 92 percent of goods declarations and 80 percent of ocean manifests are currently transmitted via EDI. Forty percent ($29.5 million daily) of customs collections of duties, taxes, and fees are electronic. Twenty-nine percent of customs cargo releases are completely paperless. The ACS computers connect to 1800 private-sector computers. The ACS database already contains 580 million records and is growing rapidly.

Productivity gains in the customs service have been evident. Since 1983, productivity has increased by over 10 percent. Considering 100 percent growth of imports during the same period, customs has increased its work force by only 7 percent. This allows it to avoid $387.4 million in additional salaries and benefits that it would have had to pay had it not realized the increased productivity. At the same time, it has enjoyed a substantial reduction in entry reject rate from 1 of every 6 in 1984 to less that 1 of 100 in 1991. The vast majority of customs brokers (82 percent) who were surveyed in 1991 believed that the Automated Business Interface (ABI) of the ACS was a marketing advantage. They also reported productivity gains and the ability to increase services to existing customers and attract new customers. With results such as these, it is no surprise that the U.S. Customs Service receives such large percentages of its data submissions electronically.

The two EDI transactions used for customs declarations and responses by the U.S. Customs Service and private-sector shippers and receivers are both EDIFACT messages. They are the customs declaration (CUSDEC) and customs response (CUSRES) messages.

EDI and the IRS

Tax administrations are particularly paper intensive operations, handling tax returns, checks, bills, notices, and other related correspondence. They must handle tax filings and payments from all corporate entities. With the advent of electronic filing and paying, business practices will change in virtually every business. The opportunities for incorporating EDI filing of data and EFT transfer of value are great. Not only can the EDI transaction sets reduce the amount of paper and speed up the process, but they can provide accurate taxpayer and tax information that will ensure correct and timely processing of tax remittances.

The current form of sharing of information using the variable-length X12 standard is not the first transfer of machine-readable information adopted by tax administrations. For many years, tax administrations have accepted data in proprietary formats on computer tapes. While these were in machine-readable format, they are currently not thought of as true EDI. In 1989, the Federation of Tax Administrators (FTA) was approached by the Tax Payments Work Group of the ANSI ASC X12 committee to participate in the development of a tax remittance advice transaction set. They were also asked by the Committee on State Taxation (COST) to champion the development of standardized guidelines for making tax payments. This became quite an issue because various states began requiring large taxpayers to remit their taxes electronically in state-specific formats. With no standard approach, these taxpayers would be forced to generate remittances and remit their taxes in up to fifty different formats. After a year's effort, the various state tax administrations, the IRS, taxpayers, and the banking community developed and endorsed a NACHA-approved tax payment convention for paying taxes. We can look forward to the growth of this EDI application as more and more companies become EDI-capable and more states require electronic filing and remittance of taxes.

International EDI

International EDI refers to the business transactions that support trade across national boundaries. Included are both corporate-to-corporate business activity and corporate-to-government activity, such as customs. Since one of the major concerns in international trade is the movement of product, the various players that provide transportation-related services such as international freight forwarders, customs brokers, and international carriers have investigated how electronic messaging might be used to streamline procedures in this industry. Their studies have repeatedly pointed to the need to reduce the high cost (sometimes almost 15 percent of the total shipping cost) of producing and maintaining accurate documentation. In addition, these studies and others have revealed that there are tremendous delays in the movement of cargo and extremely inefficient usage of transportation and port facilities, equipment, and personnel.

It became obvious that EDI messaging standards were needed to handle the specific information needs of international trade. The standard to be used for this purpose was the United Nations EDI standard for administration, commerce, and transport (EDIFACT) which will be discussed in some detail later in this chapter. In EDIFACT terminology, transaction sets are referred to as *messages.* The three initial areas of standard development to support international trade were customs, where both a declaration and a response message were developed; freight forwarding and transport to handle the actual movement of cargo; and banking, with users of the SWIFT international banking network designing messages to handle the transfer of value associated with international freight.

There are several challenges that potential users of international EDI face. One, they must adapt to the different business practices of the various nations with whom they trade. Since EDI messages pertain to the actual movement of product and since the goal is to minimize human intervention in the flow of information, it is particularly important for trading partners to be well aware of how receiving companies will interpret data and how they will act upon it. This is further complicated by the number of intermediary companies that facilitate the movement of cargo across borders. While in domestic trade, the buyer and seller are often the only parties that need to be kept in the information loop, in international trade we deal with what seems to be a cast of thousands, all needing correct and current information. It is totally impractical to attempt to impose standardization on the way businesses act in various countries and cultures. Therefore, companies entering into the international EDI arena must be aware that they must devote a good deal of time and effort into defining their assumptions, implementation steps, and anticipated actions, and then working closely with international trading partners to thoroughly test the flow of information and the related business activity that it generates.

Another challenge in international EDI is communicating across networks. Those communicating directly with trading partners through regular voice telephone lines or X.25 data networks must use the services of several national telephone companies. Often they experience time delays, lack of line security at the network hand-off points, and the need to purchase additional software to support the various interconnections. In addition, companies attempting to transmit through X.25 networks often find that these networks are incompatible with one another from one country to another and offer very different quality of service as well.

It is for these reasons that some companies choose to use EDI VANs as a clearinghouse for their international EDI messages. This solution works more effectively but is more expensive to support. And, because charges differ between EDI VANs, the most cost-effective communication solution for one trading partner may not be cost-effective for another partner. Another issue is that no one VAN has established truly international communication links with all countries, so the trading company wishing to do EDI internationally must shop around for a VAN that supports communications with the country or countries with whom it wishes to trade electronically. This often requires that companies subscribe to several EDI VANs . As VANs increase their interconnect capabilities, this issue will decline as a barrier to international trade via EDI.

Still other companies support their own private, in-house networks and choose to use them for international EDI transmissions as well. This can be an extremely expensive solution unless connectivity to the various international locations already exists and the bandwidth to handle the additional data is already available. Even so, this private network represents still another network with which international business partners must be connected.

The third challenge in international trade is to reduce the impact of different languages. Properly used, EDI lends itself well to reducing the use of language by substituting code values for narrative words and descriptions. By limiting data to codified values and using the internationally accepted EDIFACT standard for their international business messages, companies can support international business requirements. Prior to endorsement of EDIFACT, international trade using incompatible domestic standards was impossible.

Finally, there is the issue of international awareness of EDI, its capabilities, and potential benefits. It has been no mean feat to familiarize companies within one industry and within one country with the concepts of EDI and the requirements for implementing and supporting an EDI system. Internationally, this issue is magnified. As with domestic EDI, its usage will grow only if companies become aware of the benefits they can expect to realize when they adopt it as the way they do business. Spreading the word about such benefits as the following will go a long way toward building the international trading community:

- Faster communications with increased reliability and quality
- Less expensive communications
- Increased ability to accurately track shipments
- More accurate and complete business messages
- Speedier movement of shipments through customs and other administrative formalities

UN/EDIFACT Standard

Just as with domestic EDI, a generally accepted international EDI standard has paved the way for widespread implementation. The EDIFACT standard was developed as an outgrowth of work done by American and European experts starting back in 1985. They used both the existing ANSI X12 transaction set and architecture models and UN/ECE Guidelines for Trade Data Interchange (GTDI) to design the EDIFACT standard. There were two work groups formed. One worked on the overall design of the standard, while the other began design of the first EDIFACT messages: the international invoice (INVOIC) and the international purchase order (ORDERS). The latter group used the X12 invoice transaction set (810) and purchase order (850) as models.

The resulting EDIFACT standard and business messages are similar to X12 but contain differences needed to support international requirements. For example, X12 segments were modified or new segments were developed to support additional international business information requirements. Also, from the beginning, EDIFACT adopted a generic approach to the definition of data elements, while X12 had favored the specific approach. With the specific approach, each data element gets a specific name and is used for a specific purpose. For example, you might have a data element called Delivery Date or Order Date. With the generic approach, you would form a composite by linking a generic data element such as Date to a second element that acts as a qualifier code and contains one of a predefined set of valid qualifier codes. While X12 has moved toward the generic approach for data elements, it has remained rather specific when it comes to segments. In fact, as new business requirements surface, X12 segments have become longer and more complex. Some segments have upwards of 30 data elements in them. EDIFACT segments are more generic in nature. By use of a qualifier code, the intention of the segment is altered. Some segments are being developed that have very limited functionality. These are sometimes called *mini segments.* They are combined into segment groups.

Even though there are technical differences between the X12 standard and EDIFACT, the intent is to provide the same functionality in an EDI message in either standard and to allow data fields to be moved easily from one standard to the other. In fact, the plan is for transaction sets created prior to the end of 1995 to become ANSI X12 Version 4 transactions and for those created after that date to follow EDIFACT design principles and become part of the EDIFACT standard.

Currently X12 has two classifications for its EDI transaction sets. A *standard* is a transaction set whose approval has been voted for by the X12 committee and the general public. A *guideline* is a transaction set that has been approved by the X12 committee and has been released as a draft standard for trial use. The plan is to create a third category for existing X12 transaction sets not converted to EDIFACT. The status of transaction sets in this category will be higher than a guideline but not as high as a standard. By restricting the status of existing X12 standard transaction sets, unless they are made compliant with EDIFACT, the committee hopes to encourage standards developers and users to migrate existing transaction sets to compliance with EDIFACT.

From its modest beginnings to the present, the number of countries that have formally joined the UN/EDIFACT process and have formally endorsed the standard for international EDI has grown substantially. Regional EDIFACT boards that started out with few members and a narrow focus have had a swelling of their ranks and have changed their names to reflect their broader focus. For example, the Japan/Singapore EDIFACT Board changed its name to the Asia EDIFACT Board and has increased its membership with the People's Republic of China, Taiwan, Korea, Malaysia, and Hong Kong. The North American EDIFACT Board has been renamed the Pan American EDIFACT Board with the inclusion of South American countries. The Western European EDIFACT Board has had Luxembourg, Malta, and Israel join their organization. Currently, there are over 30 approved EDIFACT messages ready for use in the areas of transportation, customs, finance, and general trade. There are over 200 EDIFACT messages in various stages of design and approval.

Now let's take a look at how ocean carriers use EDI to support the import and export of product into and out of the United States. The use of EDI in this particular segment of the transportation industry is growing rapidly. In fact, there are many ocean carrier business transactions that can easily be converted to EDI to handle both import and export of goods with a large variety of transportation entities.

There are several EDI applications that deal directly with the exportation of cargo on ocean vessels. Let's look at the EDI transaction sets that support each of these applications.

First, the *booking request* also known as the *reservation request*, is transaction set 300. This transaction, sent by the freight forwarder or shipper to the ocean carrier, requests either reservation of space, a shipping container, or equipment. It contains the information needed by the ocean carrier to understand the nature and routing of the shipment and to provide any special handling requirements.

Next, the *booking confirmation* (301) responds to the booking request to confirm availability of space, containers, or equipment. This is a critical transaction set because just the sending of the request alone does not guarantee the booking. In the confirmation transaction set returned to the freight forwarder or shipper, the ocean carrier includes the booking number that it has issued. This number is used by the forwarder or shipper as a shipment reference number until the bill of lading is issued with its identifying number.

Following along the export trail, the next EDI transaction set is the *shipment information* (304). This is returned by the shipper or forwarder after receiving the booking confirmation. It contains all the information needed to generate a bill of lading and to handle freighting and scheduling. The shipment information transaction set takes the place of the shipper's letter of instruction in the paper world.

While this transaction set contains the same information as the bill of lading, it is not intended as a substitute for the paper bill. Instead it can be thought of as an electronic mechanism to facilitate generation of the paper bill of lading.

The *shipment information* transaction set is followed by the *freight details and invoice* (310). This transaction set is sent by the ocean carrier to convey information regarding shipping charges. This transaction set may be thought of as the *master document* of both the import and export chain as it contains all data related to a shipment.

Next is the *gate activity or terminal operations activity gate arrival* transaction set (322). This is sent by the shipper to inform the ocean carrier that goods have arrived at the port and are ready to be loaded into the vessel.

Sometimes the ocean carrier issues the terminal operator a *stow plan* in which it identifies the exact location into with the cargo is to be loaded and tells of any special handling requirements.

In addition, the *shipment inquiry* (313) can be sent to an ocean carrier by the shipper or forwarder to request information regarding the status or location of a shipment. The carrier may then respond with the *status details reply* (315). Sometimes the carrier transmits the reply even though no inquiry has been made. In this case the status details reply transaction set acts as a simple notification of status.

There are also several EDI transaction sets that support the import process. First is the *arrival notice* (312). This document is transmitted by the ocean carrier to the importer or customs broker to inform them that a shipment is scheduled for arrival. It contains much of the same information as contained in the paper ocean bill of lading.

Next, the carrier issues a *U.S. Customs Manifest* (309). This shows the exact contents of the vessel. It is sent to the U.S. Customs office by the ocean carrier. Here both EDIFACT and proprietary U.S. Customs–developed formats are acceptable.

Just as in the export process, the *shipment inquiry* (313) and the *status details reply* (315) are included in the import process as well. The shipment inquiry is sent by the importer or customs broker to the ocean carrier, and the status details reply is returned.

Finally, the *gate activity* (322) can be sent by the terminal operator to the carrier to inform it that the cargo has been removed from the ocean vessel and either passed to a local trucker or to a rail or motor carrier for inland delivery.

EDI-capable ocean carriers are realizing the benefits of communicating rapidly and accurately with their trading partners. As they implement more EDI transaction sets and as they communicate with a broader range of trading partners, ocean carriers can expect to attain higher levels of benefits.

To complete this discussion on international EDI, let's look at how the European Common Market intends to use EDI to facilitate its primary goals and how EDI is being incorporated into the business practices of Canadian companies and federal and provincial EDI agencies.

Uses of EDI in the European Common Market

Virginia Cram-Martos published an excellent two-part article on EDI and the European Common Market in the December 1992 issue of *EDI World*. Much of the information in this section was obtained from that article.

The main goals of the 12-member countries of the European Common Market can be described as establishing freedom of movement for goods, services, people, and capital.

Freedom of Movement of Goods, Services, People, and Capital

To attain freedom of movement for goods, member countries must remove physical barriers by suppressing or eliminating border controls. Unfortunately, this will greatly impede the collection of statistics on intercountry trade. With the elimination of customs, there will be no agency to count goods as they pass from country to country. Electronic data interchange may be able to help in this regard. Large exporters will be asked to periodically report their exports by product type and country. Reports will be requested in EDI format.

Also in regard to freedom of movement of goods are *technical barriers* of country-specific standards on all aspects of a product and the early proliferation of national and industry EDI standards such as TRADACOMS in the United Kingdom, GENCODE in the French retail industry, and VDA in the German automotive industry. This barrier would have successfully blocked the general use of EDI between European market countries. However, since the general acceptance of the UN/EDIFACT standard and general consensus on information requirements across country boundaries, this barrier is being eliminated. Likewise, endorsement of the EDIFACT standard will open Europe to the rest of the world as well.

Fiscal barriers in the form of different levels of value-added taxes (VATs) are also standing in the way of free trade between the countries of the European Common Market. Again, with the elimination of customs at the border, it will be difficult to trace intercountry shipments and the taxes owing for them. As a consequence, governments are requesting that companies supply them with information to help trace VAT requirements and payments. While not requiring this information in machine-readable EDI form, some governments will at least be allowing it. In the same way, companies who sell supplies to companies in other countries will be asked to submit reports on these supply sales. Electronic data interchange is an excellent medium for transmission of these reports as well.

In discussing the barriers to the free movement of services, one aspect is the difficulty surrounding freedom of movement of people, the service providers. To facilitate the movement of people, EDI provides timely and accurate information which may be accessed across Europe. To facilitate the performance of services across national boundaries, several industries are active. Examples are the tourism industry which has been very active in developing international EDIFACT standards; the International Air Transport Association which has developed two EDI standards; CargoIMP a proprietary system and CargoFACT based on EDIFACT; and the European construction industry which has helped to develop EDIFACT messages for the exchange of contract and purchasing information. Also, the banking industry, using the Society for World Interbank Funds Transfer (SWIFT) EDI messages for the transfer of payments internationally is looking to expand its use of EDI with more services to the full European marketplace.

Since information is so important to service industries, the structured and codified way of representing the EDI information is particularly valuable to them as a way to overcome physical, language, and cultural barriers to successful completion of intercountry business.

Uses of EDI in Canada

Similar to the United States, Canada saw EDI begin in the grocery, transportation, and warehousing industries. All used the EDI standards developed by the TDCC. Subsequently, the automotive, retail, and manufacturing industries began doing EDI but chose to endorse the X12 standard instead. For international trade, except for trade with the United States, EDIFACT is being used. Following are brief updates on industry activities.

To facilitate the flow of information regarding parts and assemblies in the automotive industry, EDI provides information as the Canada–United States border link between manufacturers and suppliers. In the retail industry, both the number of retailers and suppliers using EDI and the number of EDI transactions that they use is rapidly growing. Canadian Tire, a major retailer of automotive parts, hardlines (hardware and houseware) products, and automotive repair services, is an innovative and aggressive user of EDI. Not only does it use EDI as a technology, it also uses QR as a business strategy. To support this initiative, Canadian Tire has expanded the variety of transaction sets that it uses to include the ASN, purchase order, acknowledgment, product activity detail, payment activity detail, and payment order/remittance. In addition, the company is working with Canadian Customs to replace the existing Canada Customs proprietary CADEX transmissions with UN/EDIFACT customs documents.

Being one of the first to implement, the grocery and food industry is today still one of the largest EDI users. In this industry, many companies are branching out from the industry-specific UCS to the cross-industry ANSI X12 standard.

Many Canadian financial institutions are positioning themselves to provide flexible EDI delivery services from a close branch office. Available delivery options include fax and print as well as EDI files. Unlike the United States, which until recently restricted bank activity to one state, the banking industry in Canada is comprised of a small number of very large, countrywide banks. Countrywide banking eliminates the float issue which until recently has acted as a barrier to the implementation of EDI invoices and EFT payments in the United States. It actually facilitates the daily settling of day-to-day bank exchanges. While the United States has a national clearinghouse for interbank settlement, Canada has a bilateral exchange of credits and debits. The Canadian

Payment Association has approved the use of the X12 standard for interbank settlement of accounts.

In the health care industry, Canadian hospitals are actively using EDI for procurement of supplies. Today more than 45 hospitals and 35 of their suppliers handle procurement using CareNET, a cooperative procurement venture, founded in 1990, between the Ontario Hospital Association, the Ontario Ministry of Health, and health care facilities and their suppliers founded in 1990. They use the X12 Healthcare Industry Business Communication Council (HIBCC) subset of the purchase order and plan on expanding into the invoice, request for quote, and price sales catalog.

Canada provides an excellent example of government agencies using EDI, both on the federal and provincial levels. They are using EDI for reporting of remittance and procurement activities. In addition, Revenue Canada Customs has been using the Customs Automated Data Exchange (CADEX) system for 4 years to speed up the clearing of trains going through the Windsor-Detroit crossing and the clearing of vessels entering the ports of Halifax, Montreal, and Vancouver. These and other industry EDI applications are helping Canadian companies to stay competitive in the world marketplace.

Also helping EDI users in Canada is an organization whose goal since its inception in 1985 has been to promote EDI in Canada, the EDI Council of Canada. The council has acted as a catalyst by promoting EDI as a valuable tool for corporations and government departments to improve efficiency and lower costs. They have been spurred on by the negative impact that the continuing Canadian recession has had on businesses. Marshall Spence, president of the EDI Council of Canada, has acted as ambassador at large throughout Canada and the world to spread the word about EDI and the benefits attainable from its implementation.

Summary

This chapter has dealt with financial, government, and international uses of EDI. While there are many small budding examples, there are few full-grown programs. The potential benefits in these applications are great. However, until a critical mass is reached, it will be difficult for EDI to move from being a cost item to being a cost savings.

While it is certainly the fact that EDI will continue to grow in these arenas, the size and complexity of the applications pushes it farther down the priority list of EDI applications for American companies.

11

Futures in EDI and Electronic Commerce

In this chapter we will take a look at the various technologies that are part of the electronic commerce movement and discuss the various uses and growth potential of each. First and foremost is EDI, which, after all these years is still in the early stages of implementation in many industries and in most businesses. Considering that when EDI first came on the scene, it was touted as the panacea to cure all ills, its growth has been slower than anyone would have predicted. Electronic data interchange has proved not to be that panacea at all. Machine-readable and preformatted EDI data are most useful for intercompany messaging intended to be read and processed by a computer. While very important, it is only a small part of the larger topic of *messaging*, which refers to both intracompany and intercompany sharing of information available in many formats and on various media. In fact, businesses are realizing that where intercompany messages are intended to be used by people, the most convenient form in which to pass the information is often in narrative, print form. When those messages are graphics intended for computer processing, the appropriate format may be CAD/CAM drawings or bar-coding.

This is borne out by the Gartner Group statistic that the rate at which companies are implementing E mail with trading partners is growing 5 times as fast as the rate at which companies are adding new EDI trading partners.

Let's take a closer look at electronic commerce. What is it? Electronic commerce is the sharing of information using a wide variety of different electronic technologies, between organizations doing business with one another: customers, suppliers, banks, carriers, and government agencies. Electronic commerce also refers to the procedures, policies, and strategies required to support the incorporation of these electronic messages into the business environment.

Since electronic commerce is a fairly new concept, and since technological advances are being introduced all the time, what we consider electronic commerce today may very well be a narrow subset of what it is tomorrow. The types of messages included in electronic commerce today are

- *EDI.* Highly structured, machine-readable data meant to be generated by the sending company's computer application and processed by the receiving company's computer application with no intervening manual operations.
- *Electronic messages.* One category of electronic messaging is electronically transmitted free-form messages such as fax and E-mail. These are typically sent from one person to another but may also be generated out of a computer application. Fax and E-mail messages may be stand-alone documents or may contain additional information in the form of enclosures such as spread-sheet reports, computer files, or graphics. Another category of electronic messages is bar-coded messages, generated by a computer application and meant to be read by an electronic scanning device.
- *E forms or electronic forms.* Structured, but people-readable, preformatted messages, E forms can be used on their own or function as a user-friendly interface between the business user and a behind-the-scenes conversion to EDI.
- *Imaging.* Visual information that may be included as part of another intercompany transaction or may be a stand-alone graphic. This type of information is being incorporated into the marketing techniques used by vendors to introduce and sell their products.
- *Direct access into trading partner files.* With the focus on intercompany cooperation and closer trading partner relationships, the ultimate in service is direct access of the other organization's data files. This may be just for viewing as when a vendor reads its customer's inventory files to ascertain how much additional product it needs. It also may contain update capability to actually change the other company's data.
- *Electronic catalogs or bulletin boards.* These allow the user to merely view data or to down-load them to their own computer and print, manipulate, and/or pass them to their own computer application for processing.

As is the case with EDI, the difference between just receiving data electronically and realizing true benefits from them is the difference between a stand-alone EDI capability and one that is integrated into the system environment. The same is true for electronic commerce. Unless the various multimedia intercompany messages that a company supports are integrated with one another and truly support one another, the benefits from each can be severely limited. For example, the company that receives a bar-coded label on its incoming shipping cartons gets very little benefit from it unless it can scan the label for an automatic match to the carton identifier it received via the EDI ASN and unless it can automatically upgrade the inventory files with the contents of the carton from the same EDI transaction set. Likewise, an E-mail message can stand on its own or can reference an EDI or CAD/CAM communication and act as documentation or an alert message.

Electronic commerce should be a strategic initiative within an organization. Investments in electronic commerce can be justified according to the contribution that they can make to the company. The information services department, long considered just an internal resource to support the computer needs of the business community, is the focal point of this move to incorporate technology throughout the company. Information services can provide the wherewithal for a company to elevate its service level, improve product quality, and gain market share. The goal is to share more and more timely information with trading partners, and information services is positioned to make that happen in the most efficient and cost-effective manner.

Another aspect that surfaces with the growth of electronic commerce is the intercompany and intracompany distribution of electronic messages. Intercompany, the move is toward more robust communication capabilities able to handle several different types of electronic messaging, such as an E-mail message that can also hold EDI transaction sets and graphics or other files. The growth of X.400 and X.435 communications standards and the X.500 addressing directory can be attributed to the growth of intercompany messaging. Likewise, the ever-growing ability of companies to handle a combination of various types of electronic messages in one communication session can be attributed to the need for companies to package various electronic media into one communication envelope. Once received, electronic messages must be disseminated to people, computer programs, and sometimes machinery as in the case of manufacturing specifications.

Intracompany, corporations and government agencies are providing more and more timely information to business users throughout the organization. With database structures, everyone is given the most current view of data, and a change to those data can be seen instantly by all. With the growth of cross-functional teams, information is being shared by previously unconnected departments and business users. In order to offer widespread access to information, there has been a growth in intracompany communications linkages as well. Both local-area networks (LANs) and wide-area networks (WANs) are making it possible for all departments and divisions to gain access to the information they need. In some cases, companies have developed a long-range plan to standardize corporatewide information access across department and division boundaries. Unfortunately this is not the typical case. It is more likely that each department installs a LAN and E-mail product for its own use, with no thought to more widespread compatibility. Sometimes, a merger or acquisition has been the cause of nonstandardization. After the fact, many companies need to rethink their intracompany communication capabilities. They often install LANs to allow department users and cross-functional team members to share the information that they need and link those LANs to WANs or connect them directly to a centralized computer.

Electronic Messaging

It is predicted that E mail and mail-enabled applications will be more widely used to share business information than is EDI by 1996. With the price of E-mail services on the decline and the ability of E mail to carry structured EDI messages and other files, this seems like an easy and convenient solution for the intercompany transfer of all information. What's more, with inexpensive software packages able to convert from one E-mail format to another, this medium crosses company boundaries more conveniently as well.

Fax is also proving to be very useful, both as a stand-alone capability and as a by-product of EDI. For those cases where the sending partner is EDI-capable but the receiver is not, fax provides a print format delivery mechanism. The typical scenario is for the sender to send all its EDI messages either direct to trading partners or to an EDI VAN. When the VAN receives EDI transaction sets for a trading partner who is not EDI-capable, it translates them into print format and faxes them to the receiver. Unfortunately, this does not work well in the other direction. It is difficult and very people-intensive to translate free-form fax messages to structured EDI format.

Bar-Coding

A bar-code symbol is nothing more than a group of dark bars of varying widths separated by light spaces. The arrangement of the lines vary according to what the bar code represents. There are various bar-code symbologies used to develop bar-code symbols. The symbology is the language that contains the logic for generating the correct pattern of dark bars and light spaces. Some symbologies allow for the bar-coding of a wide variety of letters, numbers, and characters; others only allow for numbers.

Bar-code symbols are typically printed on packaging (shipment or product) or on the product itself. More recently bar-coding is also being placed on paper forms as an identifier. Bar codes are typically read through a scanning methodology. The scanner moves an illuminated spot across the symbol, picking up the dark and light band widths and placement. If the area picked up is a light space, it converts to a one or an ON electric signal; if it is a dark bar, it converts to a zero or an OFF electric signal. The quality of the printed label is extremely important to the accuracy of the scanning.

Bar-code readers (scanners) are being used to identify the bar-coded material during all stages of business. For example, bar-coded labels with bar-coded carton identifiers are automatically scanned at a receiving site. They are matched against the carton IDs in the ASN transmitted from the supplier. Based on the carton contents also contained in the ASN, the receiver can upgrade its inventory, develop stow instructions for warehouse storage, or plan placement on store shelves.

Materials management in all its many flavors provides excellent opportunities for using bar codes. Not only are bar codes useful in the receipt of product, but in its sale as well. Today, scanners are lodged in point-of-sale (POS) checkout systems. As product is sold, the scanner identifies the product. Its price is automatically added to the checkout register tape, it is automatically subtracted from available inventory, and its product code is used to begin generation of a purchase order for replenishment.

E Forms or Electronic Forms

Electronic forms are accessed on-line and are typically preformatted to look like paper forms. Electronic forms may be handled as substitutes for paper documents or they may be more or less integrated into the systems and procedures of the business. The more they are integrated, the more valuable they become. Generally, business forms are designed to accept business information through key entry and to access additional data from existing computer files. For example, a purchase order requisition form will have predesigned spaces into which the requester's name and associated information, the requested products, vendor information, and delivery information can be placed. The actual completion of the form can either be strictly manual, with the user inputting all the required fields, or it may be automated to varying degrees. For example, based on the logon ID and password of the user, all requester information can be automatically taken from a file of employees. Likewise, depending on the product being ordered, vendor information can be automatically picked up from a vendor file.

When the form is intended as a paper substitute, it is printed and handled as a paper document. As an extra value, a copy may be retained in the system for future reference. When the form is integrated into the system environment, its information may be used to populate other business transactions and as input to computer applications. For example, this same form could actually be an EDI transaction set traveling incognito for the benefit of the non–computer-oriented business user. After the form is completed, the forms program develops a computer file from its information fields. The file is fed into an EDI translator, which converts it into an EDI standard purchase order. The order is then transmitted to the product vendor via EDI. Even more system and procedural integration is possible. Information from the form may provide data regarding department, division, or corporate use of office products and may automatically be charged against the correct cost center. Depending on the product, its cost can be automatically fed into the accounting system and even automatically amortized over the correct number of years. If the price of the purchase exceeds the allowable dollar value of the requester, the form can automatically be routed to a manager's computer. Once authorized, the manager's authorization can be associated with the order as well.

This same scenario can be used when data are received in the highly structured machine-readable EDI format but are destined for a business person. The EDI data fields can be converted from codes to words and expanded from identifiers to descriptive information and placed into the predefined spaces in the form. The completed form is then made accessible to the business user's microcomputer.

Today, forms program vendors are developing alliance partner relationships with business application vendors, for the purpose of integrating forms with business systems; and with EDI translator vendors, for the purpose of integrating forms with EDI.

Imaging

A picture really is worth a thousand words! Companies are finding all sorts of innovative ways to augment transmitted data and other computer-oriented information with graphics and photographs just as we have done for years with paper. Take for example, the on-line catalog that contains photos of actual products for sale. There are a host of applications growing up around the ability to transmit visual information and the incredible growth of storage and processing power on microcomputers.

Imaging-related applications can be more or less integrated with other systems. In home centers today, customers are able to consult with kitchen specialists who help them to develop remodeling plans for their homes. Today, the system they use is initiated by a floppy disk that contains information specific to one brand of kitchen cabinets. To look at another brand, another floppy starts a new program. What if the microcomputer contained a catalog of kitchen cabinets from all vendors? Once the dimensions and door and window placements are defined to the system, customers could view any cabinet lines in which they are interested to see how they would look in their own kitchens. At the same time, they could see the projected price of the total order. Once the selection is made, the identifying numbers of the cabinets being ordered could be used to automatically generate an EDI order which could then be transmitted to the furniture manufacturer. The same information could be used for sales analysis reports.

Another example of this type of system is a deck-planning program. Here the customer could select the deck that he or she would like to build. The system would calculate the wood requirements, generate and transmit the order, and even plan the truck delivery through EDI transactions behind the scenes.

Also in imaging technology is character recognition. In this scenario, a business user would write information in words. The words would then be scanned and used as input to EDI transaction sets. Again, with more integration into the systems area, scanned free-form data could be cross-referenced to files containing codes and identifiers which would be used to reduce the amount of data needing to be transmitted. For example, company name and address could be cross-referenced to a customer or vendor ID number. Contact name could be used to access additional contact information such as a telephone number.

Direct Access into a Trading Partner's File or Database

Even though there are rapid and secure technologies in place to transmit data, the most efficient and time-effective method of gaining access to a trading partner's data is to directly access its computer files. In the same way that trading partners are developing closer business relationships, they are beginning to share information through this type of direct access. This has long been the case in Japan, where the major vendors of a large manufacturer have access to production schedules and inventory files of their customer.

Electronic Catalogs and Bulletin Boards

Electronic catalogs provide accurate, up-to-the-minute information in a form that can be used as input to computer applications. An example of a popular electronic catalog is the product code catalog. Data for this catalog are provided by the vendor via an EDI transaction set. Each product manufactured by the vendor is listed by identifying UPC number and other pertinent descriptive information. The customer accesses the catalog entries and gets the correct UPC number to place in its order. As you can imagine, the complete list of all UPC-numbered products is huge. Even a list of just those from a specific vendor is tremendous. To reduce costs and facilitate the access of catalog information, some outside organizations have put the UPC catalog on-line. Vendors keep the catalog current by sending their latest UPC information to the catalog company on a regular schedule. Customers send in EDI transactions that request categories of specific codes to be returned to them. They try to zero in on just the product line in which they are interested. For example, they may request UPC listings of Nike shoes for women as opposed to the full line of Nike shoes. The public UPC catalog was developed to service apparel vendors and retailers. It is used regularly by the retailers and these vendors. However, there are much broader applications for this catalog. Other retailers, such as the major home centers, are extremely interested in having their vendors put UPC information on the catalog as well.

Bulletin board services are actually penetrating into small companies. They offer an inexpensive way of keeping track of what other companies are doing. Bulletin boards can be used to develop person-to-person interactions when companies do not regularly travel to see one another. For example, companies can run interactive electronic meetings on bulletin board services. In addition, bulletin board services can be repositories of information to augment business. One trading partner can put information on the service for all its trading partners to access. Software vendors can put product updates on the service. Even the latest version of the EDI standard and articles related to business can be accessed and down-loaded from the service.

In addition, corporate-to-corporate sales may very well be moving toward this type of bulletin board approach. Not only would it be more convenient to shop in this way, but prices would be substantially lowered by eliminating the costly person-to-person sales call.

Finally, the government solicits vendor bids by placing product requirements on a bulletin board. This is an inexpensive way for perspective suppliers to the government to learn of the purchases various government agencies are planning.

Summary

We've discussed several of the possible components to an electronic commerce capability. Which ones a company chooses to use should be a function of what its business needs are. The most effective way of selecting components is to analyze how each technology would interact with both systems and people to respond to the current business need. The more technology independent you are, the more you will avoid tying yourself into solutions that are less than efficient to solve your current problem and the more flexible you can remain. It has been estimated by Gartner Group that 30 percent of large enterprises will implement some form of technology-independent EDI strategy by 1995. That number goes up to 50 percent by 1997.

Appendix A: Lists of Sample Tasks by Stage

Note: The following samples are certainly not meant to be exhaustive lists of tasks and make no claims to be appropriate for your business. They are provided just as a sample to show where certain types of tasks are placed within the five stages of a paper document's life cycle.

For Incoming Paper Document

Stage 1: Receive and Distribute

- Retrieve mail.
- Sort out incoming purchase orders.
- Deliver to customer service for order entry.
- Count and batch incoming purchase orders by customer service representative.
- Deliver batched orders to customer service representative's desk.
- Note date, time, and number of orders given to each representative.

Stage 2: Initial Processing

- Count number of orders in batch.
- For each customer name, look up and note customer ID number on order.
- Sort orders by customer ID number.
- For each order:
  1. Read header information.
     a. If ship-to information is missing, look up default ship-to information in customer profile report.

   b. If date is missing, call the customer to ascertain correct order date.
2. Check requested ship date.
   a. If date is missing, enter default date as 1 week later than current date.
   b. If date is less than 4 working days from current date, look up truck route to see if requested date is possible. If so, mark order as rush order. If not, call customer and tell them the soonest date product can be delivered.
3. Check shipping preferences against the list of trucking or rail companies available to deliver product.
4. Validate the bill-to information by comparing to the customer profile report. If information is different than in the company file, call the customer and ascertain correct information.
5. For each purchase order line item:
   a. Check product code against company's list of valid product codes. If not found, look for product code in customer profile list of customer product codes.
      (1) If found, cross-reference customer product code with company's and substitute for company's in the purchase order.
      (2) If not found, call the customer and get the correct product code.
   b. Compare the customer's product description with the company's for the product code. If different, check with the customer to be sure it has sent the correct product code.
   c. Make sure that there is a numerical value in quantity.
   d. Check quantity/unit of measure (UOM) for reasonableness. For example, is 144 dozens correct or should it be 12 dozen. If suspicious, call customer and verify correct quantity/UOM.
   e. Compare line item price with product and price file. Check quantity discount schedule. Check customer profile information for special deals.
   f. If a ship-to address is found in a line item, separate this line item into a new purchase order with this delivery address before entering.
   g. Check to see if this product purchase is taxable or part of the customer's cost of production.
6. Compute total purchase order cost. If different than incoming order, call the customer to apprise it of the variation.
7. Enter all purchase order information into the order-entry system.
8. Place the customer service representative's initials in the appropriate field.

9. When a batch is completed, return paper orders to the supervisor's desk and sign off on completion time and date.

Stage 3: Filing, Distributing, and Accessing Document Information

- File completed orders by customer ID number.
- Handle purchase order inquiries by retrieving completed order from file cabinet.
    1. If not found, check through batches on supervisor's desk awaiting processing or check through batches on other customer service representatives' desks.
    2. If found, relate appropriate status to customer—such as awaiting processing, entered and awaiting staging, already picked up by transportation company, or will arrive early/late/on-time.
- Handle changes to orders.
- Write requested change (product code, quantity, UOM, requested delivery date, etc.) on paper purchase order along with requester's name, date, and representative's initials.
- Send copy of changed purchase order to the warehouse for an inventory check and staging of new order.

Stage 4: Reconciling

- Check order against inventory file to ascertain that product is available to fill it.
    1. If so, check on warehouse location of stock on warehouse report, develop picking slips for product, degrade available inventory in inventory file, and write actual shipping quantities on copy of order.
    2. If not, write up back order and call customer to tell when shipment will be received.
- Send copy of purchase order with actual shipment quantities to accounts receivable department for generation of invoice.
- Save copy of purchase order with actual shipment quantities for monthly inventory count reconciliation.
- Send copy of purchase order with actual shipment quantities to accounting for filing.

Stage 5: Historical Filing and Ongoing Accessing of Paper Document Information

- File copy of purchase order in accounting file cabinet.
- Access final copy of order if customer calls after product has been delivered, invoiced, and paid for.
- Access copy of order for year-end audit.
- Send copy of order to off-site storage for IRS files.
- Send mailroom clerk off-site to retrieve IRS copy of order when long-term question arises.

Appendix B: Developing an Information-Flow Diagram

Information-Flow Diagram Symbols and Their Meanings

The four symbols illustrated in Fig. B-1 are used to depict activity in a specified area of your company, for example, sales, purchasing, accounts receivable, accounts payable, traffic/distribution.

Use the *square* to represent an entity outside of the scope of this flow, such as a trading partner or a company employee in another department or division of your own company.

Use the *circle* to represent a task or group of tasks. The task should be described with an active verb and may be handled either by a system or manually. Examples of system tasks are

Run order-entry application.

Check inventory availability.

Examples of manual tasks are

Distribute orders to correct order-entry clerk.

Validate ship-to information.

Use the *rectangle* to represent data stores, collections of information. A data store may be a paper report, list, catalog, etc., or it may be a computer file or database.

Use the *arrow* between the other symbols to represent the flow of information between them. In business situations, information from a data store is pulled into a process to complete a task, is put out to a data store as output of the task, or is pulled in, updated, and then put out again to the same data store or to another.

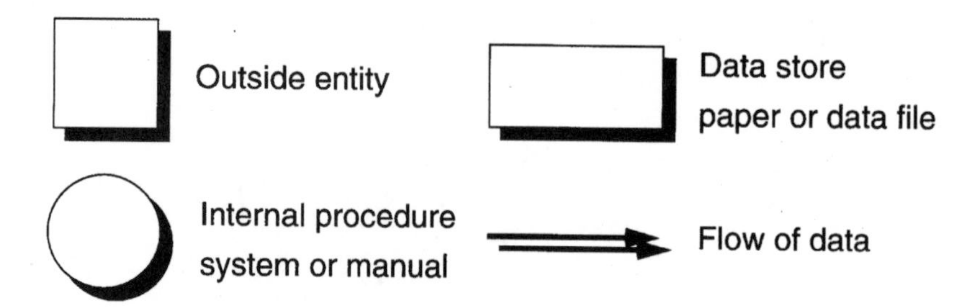

Figure B-1. Symbols used in information and data-flow diagrams.

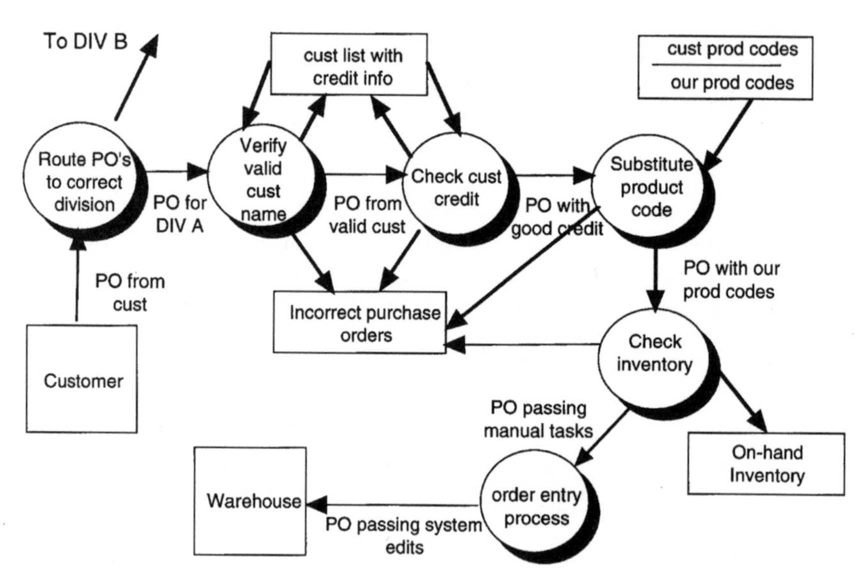

Figure B-2. Sample information flow diagram showing incoming purchase orders and associated functional area tasks and systems.

Figure B-2 illustrates a very simplified information flow showing

- Purchase orders arriving into the sales area of a company
- Those for division A being processed, those for division B being directed to that division for processing
- Manual edits on customer name and product codes being performed
- Running of a computerized order-entry application

- Good, complete orders, passing all manual and system edits routed to the warehouse for fulfillment
- Processing exceptions, those failing any of the manual edits, placed into an incorrect purchase order data store to be handled by a person

If you find that you have too many circles (more than 10) in your information flow, try grouping tasks. For example, instead of having a separate circle for each edit done on the purchase order header information, have one that says "Edit PO header fields." In a separate diagram entitled "Edit PO header fields" you can show all the processes, data stores, and information flows that comprise that one circle. So, in effect you are developing a hierarchy of information-flow diagrams, going from the least to the most detailed.

Glossary

Accounts payable: a business area within a company into which the invoice is typically routed and from which the responding payment is generated.

Accounts receivable: a business area within a company that generates an invoice for products or services sold by the organization and receives the responding payments.

ACH: an automated clearinghouse—through which banking transactions affecting more than one financial institution are routed, to debit and credit the correct financial institutions.

ACH credit: a transfer of funds from buyer to seller bank account initiated by the request of the buyer.

ACH debit: a transfer of funds from buyer to seller bank account initiated by the request of the seller.

ACK: in communication, the positive acknowledgment that the receiving error-detecting communications protocol returns to the sending side to say that it has accepted the previously transmitted block of data. A NAK is the negative acknowledgment that is returned to say that an error has been detected and the previous block must be retransmitted.

Ad hoc: on request for a special purpose.

Advance shipping notice (ASN): an EDI standard transaction set that is transmitted from a supplier to its customer to describe the products and quantities in a pending shipment and the anticipated time of its delivery to the customer site. Also called the ship notice/manifest.

AIAG: the Automobile Industry Action Group—a committee that designs EDI transaction sets and handles standard maintenance requests for the automotive industry. Also the name of the industry guidelines for use of the X12 standard and the name of the subset of the X12 standard used by the automotive industry as well.

AISI: American Iron and Steel Institute.

ANSI: the American National Standards Institute—an organization that maintains standards on many different topics.

ANSI ASC X12: the American National Standards Institute, Accredited Subcommittee X12—the ANSI subcommittee that has been charged with developing the architecture and syntax rules for the variable-length X12 standard and the format and content for all business transactions that can be converted to EDI. The mandate for all X12 transaction sets is that they have cross-industry application.

ANSI X12 data element dictionary: the X12 document that contains definitions and attributes of all data elements used in ANSI X12 transaction sets or in any of the standard control envelopes. This dictionary also contains the list of valid code values associated with each element that uses a code.

ANSI X12 segment directory: the X12 document that contains name, description, and schematic diagram of each segment used in ANSI X12 transaction sets or in any of the standard control envelopes.

API: American Paper Institute.

Application acknowledgment: EDI standard transaction set (824) sent by the receiver of EDI transaction sets to the sender to acknowledge of receipt and processing by a business application.

Application format/file: the specific placement of fields in a file that is generated by a computer program or expected as input to a computer program.

Application link (software): a program that acts as a link between a business application program and an EDI translator. For an EDI sender, the application link accepts data from a business application and generates the file needed by the EDI translator. For an EDI receiver, the application link accepts data from an EDI translator and generates the application file format needed by the business application.

Application-to-application: the use of data generated by computer program directly by another computer program with no intervening manual tasks. In EDI, these programs are housed at the different trading partner sites.

Architecture (of an EDI standard): the form and format of control and transaction set elements of a standard. In the public EDI standards such as ANSI X12, EDIFACT, UCS, WINS, TDCC, there are set rules that specify how segments, transaction sets, functional groups, and interchanges must be represented to adhere to the architecture of the standard.

Arrival notice: an EDI standard transaction set sent by an ocean carrier to the subsequent transportation carrier such as an inland trucker and to the consignee (receiver or customer) to report on the arrival of a shipment.

ASC X12: see ANSI ASC X12.

ASN: see Advance Shipping Notice.

Asynchronous protocol: the software that controls the transmission of data where each character is preceded by a start bit and followed by a stop bit and intervals vary between characters.

Asynchronous transmission: the movement of data across communications lines controlled by a synchronous protocol.

Audit trail: the record of transactions generated by tracking the successful completion of various computer processes or manual procedures.

Authentication: the procedure to ensure that any tampering with data that occurs during transmission is detectable by the receiver. The EDI sender generates, through a series of complex calculations initiated by a secret key (authentication key), a macroauthentication code (MAC). The MAC is appended to the EDI data stream and the full stream is transmitted to the receiver. Upon receipt, the identical calculations are performed, initiated by the same authentication key. If the data have been changed in any way, the MAC computed at the receiving site will be different than that sent.

Authentication key: a string of characters which is used to initiate the authentication process. Each partner of a trading partner pair must have possession of the same key.

Autodial: a facility of an EDI VAN which dials the appropriate number and transmits EDI data to a recipient company or another EDI VAN either immediately upon receiving them or at a subsequent time based on a predefined schedule or on volume of data awaiting transmission.

Automated environment: a business environment in which tasks are handled by computer programs with no need for human intervention except for the handling of exceptions.

Award bid notification: an ORDERNET/NWDA standard transaction used primarily by the pharmaceutical industry. A manufacturer sends the award bid notification to a distributor to describe details of a special contract between itself and a hospital.

Back-end processor: a small computer on which some processing is performed that follows the main application processing. Usually attached to a larger computer on which the main application processing is performed, this is the computer on which subsequent processing is done after completion of initial processing on another computer.

Bandwidth: the capacity of a telephone line to carry data. The higher the bandwidth, the more data that can be carried at one time.

Bar code: an array of parallel dark rectangular bars and white spaces that together represent a single data element or series of characters in a particular symbology.

Baud rate: the transmission rate, used as a measure of serial data flow between computers or between communication equipment devices.

Best practices: in the grocery industry, the most efficient and cost-effective ways for a company to handle business transactions and specific programs.

Bid award notification: see Award Bid Notification.

Billing transaction: a data stream containing information relating to charges for products or services rendered.

Bill of lading (BOL): in EDI, an EDI standard transaction set sent by a shipper to a carrier that describes shipment contents and delivery details of an order awaiting pickup.

Bisynchronous protocol: software that controls the synchronous transmission of data. The protocol has predefined start and end characters and error-checking routines.

Bits per second: the number of bits that pass a given point in a communication line per second.

Blind entry: in the direct store delivery segment of the grocery industry, this refers to acceptance of and upgrade of inventory information for product delivered by scanning the bar code of the product and accepting quantity from the invoice with no manual counting of product units.

Block of data: a predefined number of characters that comprise the unit called a block.

Booking confirmation: an EDI standard transaction set that contains information to reflect the acceptance of a booking for freight on an ocean carrier.

Booking request: an EDI standard transaction set which contains information to reflect the shippers desire to place freight on an ocean carrier.

Book transfer: in the banking industry, the recording of a transfer of funds from one bank account to another in the same financial institution.

Bulletin board: an interactive system where various users can place and retrieve information that is pertinent to a particular topic area. Bulletin boards often have question-and-answer forums as well, where users can ask and answer questions on the specified topic area.

Business application: a program or group of computer programs that perform related tasks.

Business trigger: see Trigger.

CAD/CAM: Computer-aided design/computer-aided manufacturing—machine readable data able to be read, interpreted, and acted upon by a computer program.

Carbon copy routing: the transmission of duplicate EDI standard transaction sets to one or more additional receiving locations or trading partners.

Cash management: the process of controlling cash by keeping track of accounts payable and receivable, making financial decisions that maximize interest earned by accelerating receipts and holding disbursements until the due date.

Cash position: the comparison of the dollar value of a company's payables and receivables.

Catalog: see Electronic Catalog.

Catalog management: in the grocery industry, the management of inventory levels, replenishment, placement, and marketing of a group of related products, as opposed to the more traditional handling of each product individually.

CCD: cash concentration and disbursement—in banking, an electronically transmitted ACH standard transaction set for transferring funds from one account to another for purposes of making a payment.

CCD+: a CCD transaction set with addenda records in which narrative-style comments on a payment can be placed.

Central bank: a commercial banker's bank. It accepts deposits, makes payments, and extends loans to commercial banks.

Centralized ordering: an ordering or procurement department that provides ordering services to most or all departments and/or divisions within an organization.

Centralized processing: a system that handles data at a single location within an organization. It either accepts input from various departments and divisions or subsequently routes data to various business users.

Charge back (debit memo): the ORDERNET/NWDA standard transaction set that is sent from pharmaceutical distributor to manufacturer to request a rebate based on the sale of product by the distributor to a hospital that has contracted for a special price with the manufacturer.

Charge-back reconciliation: the ORDERNET/NWDA standard transaction set returned by a pharmaceutical manufacturer to a distributor to relate corrections to the charge-back debit memo. It is typically generated out of a reconciliation procedure that compares the charge back with the hospital and manufacturer contract terms.

CIDX: Chemical Industry Data Exchange—a group formed to coordinate EDI activities for the chemical industry. Also the name of the guidelines for usage of the X12 standard for the chemical industry.

Codify: the replacement of narrative-style data by predefined code values.

COMM ID: an identifier used by grocery industry senders and receivers of EDI data to identify themselves to their EDI trading partners.

Communication board: an electrical panel which allows a computer to perform communications via telephone line.

Communication capability or communication configuration: the protocol and line speed at which the communication device component of a computer system is able to transfer data to another computer or communication device.

Communication medium: the technology used to communicate data.

Communication protocol (software): see Protocol.

Communication software: the logic and computer instructions that monitor and control communication activities.

Compatibility: the ability of computers, communication devices, and software programs to accept and process data prepared by another computer, communication device, or software program without conversion or code modification.

Competitive edge: a favorable position over competitors.

Computer application/business system: the computer program which solves a problem or performs a function. Examples: the order-entry application handles incoming purchase orders; the billing application generates invoices.

Computer memory: the space within the computer in which data are stored and from which data are accessed.

Computer operator: the person who actually manipulates computer controls.

Computer pad: a device attached to a computer on which graphical information can be drawn, for example, a signature; and which can transmit that data into a computer application. The computer pad provides a means to get otherwise non–machine-readable information into a computer file.

Computer program: a set of logic and instructions for solving a problem with a computer.

Computer storage: the portion of the computer that receives data, retains them for an indefinite period of time, and supplies them upon command.

Computer to computer: the transference of data generated by one computer to another computer.

Conditional usage designator: see Usage Designator.

Control envelope: a pair of segments that mark the beginning and end of an interchange, functional group, or transaction set in an EDI standard. A control envelope header contains information needed to identify the entity to follow. The trailer contains information to match the trailer to the associated header and to provide the count of the number of units in the entity. Control header information is generated by the sender of EDI data and is used by the receiver to verify that the complete entity has been received.

Control number/identifier: the number used to identify an entity in an EDI standard. For example, a segment identifier identifies a standard segment, a data element identifier identifies a standard data element, and an interchange identifier identifies an interchange.

Control totals: an EDI standard transaction set sent by a buyer to its bank to provide totals for one or more remittance advices or sent by a bank to its customer to provide totals for one or more lockbox transactions.

Corporate trade exchange: see CTX.

CPU: central processing unit—the central processor of a computer system, containing the main storage and arithmetic computation unit.

Cross-docking: the immediate movement of product received at the dock of a warehouse or distribution center to the dock from which it will be transported to the subsequent receiving location. By arranging for immediate cross-docking of incoming product, retailers are able to reduce to a minimum in-transit time for their incoming product.

Cross-industry standard: an EDI standard that has cross-industry application. By mandate of the ANSI organization, X12 is such a standard. Broad participation by many industries in the design of the standard transaction sets and public review and acceptance prior to becoming a standard ensures that X12 transaction sets fulfill the data needs of many industries.

CRT: cathode ray tube—computer terminal that may be used for display only or may be fully programmable.

CTP: corporate trade payment—a system of handling electronic funds transfer, usually through the use of an ACH debit.

CTX: corporate trade exchange—the process of effecting electronic funds transfer through use of the payment remittance (820) EDI transaction set and an ACH credit.

CUSDEC: customs declaration—an EDIFACT standard message sent by an importer to the customs department to describe product that it is importing.

CUSRES: customs response—an EDIFACT standard message sent by the customs department to an importer that responds to the CUSDEC transaction set.

Customer account analysis: an EDI standard transaction set sent by a bank to its customer to report on account balances and/or financial transactions.

Customer identifier (ID): a code that identifies a customer organization in the interchange segment of an EDI data stream.

Data: a general term to denote the basic elements of information which can be processed or produced by a computer.

Data availability: those data elements that are available from internal company files or as output of a computer application for purposes of generating an EDI data stream.

Database: vast and continuously updatable file of information on a particular subject. Usually a database can be randomly accessed by a search of the values in predefined key fields.

Data communication: the transmission of information from one point to another.

Data element or data field: a group of characters that specify an item at or near the basic level. The smallest unit of data defined in an EDI standard.

Data element delimiter character: a character that marks the end of information contained in a variable-length data field.

Data element number: see Data element reference number.

Data element reference number: the number which identifies a data element in an EDI standard with its corresponding definition in the data element dictionary.

Data element usage indicator: see Usage designator.

Data flow diagram: a visual representation of the flow of data throughout the routines of one or several computer programs.

Data-flow path: the route taken by a specific data element at each step in a system or in a business procedure.

Data integrity: in EDI, the maintenance by a receiver of data that are identical to those transmitted by a sender.

Data requirements: refers to those data elements needed by a computer application in order to process a specified business transaction.

Data stream: all data transmitted through a channel in a single transmission.

Data type: the characteristic of a data element that describes whether it is numeric, alphabetic, or alphanumeric.

Deauthentication: actually reauthentication, performing the same authentication process at the receiving side of a transmission as that performed at the sending side. This process must be initiated with the same authentication key as that used by the sender. If the receiver computes the same MAC as was computed by the sender, the data stream can be assumed to be identical to that sent.

Debit authorization: an EDI standard transaction set that expresses the preapproval by a buyer to its financial institution to debit its account to make a payment for product or services that it has purchased.

Debug: to locate and correct any errors in a computer program.

Decryption: the reversal of an encryption procedure in order to unscramble or decipher an encoded data stream.

De facto: in fact, in reality.

Demographic information: information regarding vital and social statistics, for example, from the study of populations and their habits.

DEX/UCS: direct exchange of UCS data—refers to the EDI standard transaction sets used by the direct store delivery segment of the grocery industry; specifically those transaction sets carried by the supplier agent (route driver) to the back door receiving computer of the retailer.

Dialyzer: an apparatus containing a semipermeable membrane that acts as a substitute for a kidney and filters waste products from the blood.

Dialyzer cell: the component of a dialyzer apparatus that actually does the filtering of waste products from the blood.

Direct benefits: benefits directly attributable to the implementation of EDI. Such benefits as reducing staff and eliminating the need for key entry and paper forms are examples of direct benefits of EDI.

Direct store delivery (DSD): refers to the delivery of goods to a retailer where the seller, either manufacturer or distributor, has responsibility for the replenishment of stock on the shelves and performs delivery of product toward that end.

DISA: Data Interchange Standards Association—the organization responsible for performing all administrative duties of the ANSI X12 subcommittee.

Discount terms: refers to the discount offered by a vendor to its customer if an invoice is paid within a specified period of time.

Disk: the circular metal plate component of a computer on which data are stored; it continually rotates to allow data to be read from it and data to be written to it.

Distributed processing: having multiple sites and/or computers within a company that perform the same type of processing. For example, an organization may generate outgoing purchase orders from various departments or divisions and may likewise process incoming invoices in various departments and divisions.

DOD: the Department of Defense.

Down-load: the transfer of data from a large computer to a front- or back-end processor (smaller computer).

Down time: the period during which a computer is malfunctioning or not operating because of regular maintenance or machine failures.

Drop-ship: in the retail industry, refers to the delivery of product at its final destination point, usually the retail store.

DSD: see Direct Store Delivery.

DSD/UCS: direct store delivery use of EDI data—refers to the EDI standard transaction sets used by the direct store delivery segment of the grocery industry.

DUNS number: the number assigned by Dun and Bradstreet to identify a corporate entity. The DUNS number is often used by EDI senders and receivers to identify themselves in the interchange header of an EDI transmission.

EAGLE: a fixed-length EDI standard developed for and used by wholesalers, retailers, and manufacturers in the hardware and houseware industry.

EC: electronic commerce—the sharing of information via various electronic means such as EDI, E mail, fax, file transfer, transmission of CAD/CAM drawings, both intracompany (between different departments and divisions) and intercompany (between trading partners).

ECR: efficient consumer response—combination of technologies and business procedures and practices that eliminates waste and inefficiencies throughout the supply chain in the grocery industry.

EDI: electronic data interchange—intercompany, computer-to-computer communication of data which permits the receiver to perform the function of a standard business transaction and is in a predefined standard data format.

EDIA: EDI Association—formerly the TDCC, an organization that developed and maintained EDI transaction sets for the transportation, grocery, and warehousing industry.

EDI data: a data stream that complies with the architecture and syntax rules of an EDI standard.

EDI/EFT: those EDI standard transaction sets used to effect electronic transfers of funds from one bank account to another.

EDIFACT: EDI standard for administration, commerce, and transport. This standard contains data requirements for carrying on international trade and has been accepted by countries all over the world as their EDI standard.

EDI gateway: in EDI, the functional business area that is the logical entry or exit point for EDI data coming into or being transmitted out of a company. The EDI gateway for incoming purchase orders is the sales department; for outgoing purchase orders it is the purchasing department. In communications, it is a device having output and input channels through which communication of EDI data is effected.

EDI potential: an indicator of the potential payoff of implementing EDI; derived by subtracting the setup and ongoing costs of implementing and performing EDI from anticipated short- and long-term savings and other benefits.

EDI rating: a measure of the benefits that can be realized by conversion of a business document to EDI and automating its processing.

EDI readiness: an indicator to measure a firm's ability to physically accomplish and attain benefits from EDI based on the presence of computer and communication hardware and application software.

EDI relationship: the business relationship between trading partners that share information via EDI.

EDI standard: see Standard data format.

EDI standard file: a machine-readable file that complies with the architecture and syntax rules of an EDI standard.

EDI system: computer programs that act as the link between a business application and transmission of EDI data; typically comprised of a communication capability, an EDI translator, and an application link program.

EDI transaction (set): the electronic representation of a business document. Comprised of a predefined list of EDI standard segments starting with a transaction set header and ending with a transaction set trailer. Most EDI transaction sets have three sections, a heading section for data fields that refer to the whole transaction set, a detail section for data fields referring to individual line items, and a summary section for totaling data.

EDI translation software: see Translation Software.

EDI VAN: a value-added network that offers protocol and line speed matching to facilitate communication between trading partners and other EDI-related services such as carbon copy service, conversion of EDI transaction sets to fax or mail, and trading partner implementation services.

EDX: electrical industry data exchange—a group formed to develop EDI data requirements for the electrical industry and to represent the industry with maintenance requests regarding the X12 EDI standard.

E form: see Electronic form.

EFT: electronic funds transfer—computerized systems that process financial transactions and information about financial transactions; specifically, the exchange of value between two financial institutions.

EIDX: electronics industry data exchange—a group formed to develop EDI data requirements for the electronics industry and to represent the industry with maintenance requests regarding the X12 EDI standard.

Elastic demand: a demand that changes in relatively large volume in response to an increase or decrease in price and in the opposite direction. A small upward movement in price elicits a big reduction in demand

Electronic catalog: a database of information describing a predefined group of products that can be accessed electronically by either interactive access to the catalog or by request and retrieval of information through file transfer. An example of an electronic catalog is the UPC catalog containing product codes and related product information of the product lines of those vendors who participate in the catalog.

Electronic commerce: see EC.

Electronic form: usually an on-line representation of a paper form that can be accessed and completed in an on-line, interactive session and is sometimes processed by a computer system as well.

Electronic funds transfer: see EFT.

Electronic invoice: invoice transaction in standard data format.

Electronic mail: E mail—a system that provides the ability to generate and deliver user-generated, human-readable messages to any and all defined locations (addresses on the system). E mail may be an intracompany capability or it may be used between companies.

Electronic mailbox: see Mailbox.

Electronic trade: see EC (electronic commerce).

Electronic transmission: sending of data via any sort of electronic means.

E mail: see Electronic Mail.

Encryption: the procedure of scrambling data by performing a series of calculations initiated with a secret key. The resulting encrypted data stream is indecipherable to the outside observer. The receiver of encrypted data must decrypt the data, performing the same calculations initiated by the same key to convert the data back to its original form.

Encryption key: a string of characters used to initiate the encryption process. Each partner of a trading partner pair must use the same key to initiate the encryption-decryption process.

Error-detecting (protocol): a protocol having the logic to detect when data received is not identical to that sent.

Ethical: in the pharmaceutical industry, refers to a drug that is sold only with a physician's prescription.

European Common Market: a group of several European countries that intend to eliminate barriers for trade between themselves, by eliminating tariffs and customs checks and reducing paperwork related to the intercountry buying and selling of product.

Evaluated receipts settlement (ERS): a procedure for authorizing payment for product that compares anticipated receipts to actual receipts and pays on actual receipts with no invoice. This procedure is used extensively in the automotive industry where prices have already been agreed to in a blanket purchase order.

Exceptions: error conditions—refers to any instance of an error condition such as when invalid or incomplete data are found in an EDI transaction set during a manual or system procedure.

Express mail: refers to a service offered by a third-party service provider (VAN) in which EDI data are translated into print format and mailed to a trading partner recipient.

Fax: facsimile—transmission of text, pictures, maps, diagrams, etc., by radio waves. The image is scanned at the transmitter and reconstructed at the receiving station.

Feasibility stage: the stage during which the potential benefits and costs of implementing EDI are evaluated. The deliverable of this stage is a feasibility study.

Feasibility study: a document that reports on the outcome of investigation into the business and systems environment of a company and into outside economic conditions, competitive situation, and external pressure from trading partners. A feasibility study is used to build the case for EDI and sell EDI and its accompanying costs and benefits to corporate management. Once accepted, the feasibility study becomes an integral part of the corporate implementation plan.

Fed Wire: a method of transferring funds from one account to another or from an account to cash via a banking transaction that is wired from one financial institution to another.

Financial EDI: refers to EDI activity that supports billing, paying, and handling financial reconciliations and reporting.

Financial information reporting: an EDI standard transaction set sent by a bank to its customer to report on account balances and/or financial transactions.

Financial institution: a private or governmental organization which serves the purpose of accumulating funds from savers and channeling them to individuals, households, and businesses needing credit.

Financial return notice: an EDI standard transaction set sent by a financial institution to its customer to report on items returned by the ACH system.

Financial transaction: a data stream containing information relating to payment or any other financial application within a company.

FIPS: Federal Information Processing Standard.

First-tier supplier: the direct supplier of a manufacturer or assembler organization.

Fixed-length file: a machine-readable file where each data field and each record are of a predefined fixed length.

Format syntax rules: see Standard syntax rules.

Free-form: human-readable words and sentences, not codified.

Freight bill: a bill generated by a carrier to cover the costs of carrying a shipment.

Freight details and invoice transaction: an EDI standard transaction set that relates information regarding the carrying of freight and its related cost.

Front-end processor: a small computer on which some processing is performed that precedes the main application processing. Usually attached to a larger computer on which the main application processing is performed.

Functional acknowledgment (FA): an ANSI X12 transaction set generated by the receiver of EDI data and transmitted to the sender. The functional acknowledgment reports on both receipt of the data and the results of validation edits. It may be produced and report on the functional group, transaction set, or data element level.

Functional (business) area: a part of a company that is charged with performing specified tasks or accomplishing specified functions.

Examples: sales, marketing, accounts payable, accounts receivable, purchasing, distribution.

Functional business manager: manager of a functional business area.

Functional group: a group of like EDI transaction sets in an EDI standard file.

Functional group envelope: a pair of EDI standard segments that mark the beginning and end of a standard functional group.

Functional group header segment: a segment that marks the beginning of a functional group. It contains group type, group identifier, sender and receiver IDs, and version of the standard used.

Functional group level: refers to the receiving, sending, or processing of a full functional group.

Functional group level acknowledgment: a functional acknowledgment which acknowledges receipt of data and reports back to the sender on the completeness and correctness of the functional group.

Functional group trailer segment: a segment that marks the end of a functional group. It contains the same group identifier as contained in the associated functional group header and the number of transaction sets contained within the group.

Gate activity or terminal operations activity gate arrival: EDI transaction set that contains information regarding arrival of a shipment.

Gateway (computer) application: a computer program or group of programs that receives EDI data directly after being translated from the EDI standard data or that produces data to be translated into the EDI standard. The first business system that EDI data are used in at the receiving site or the last business system performed before converting internal data fields into an EDI data stream at the sending site. For example, the order-entry application is the gateway application for incoming purchase orders; the purchasing application is the gateway application for outgoing purchase orders.

Generic standard: an EDI standard with cross-industry application such as the ANSI X12 standard.

Group level acknowledgment: see Functional group level acknowledgment.

Guideline: in the ANSI X12 EDI standard, a transaction set that has been approved by the membership of X12 but has not yet gone out for and passed the public review process.

Guidelines for usage: a document that describes the data needs of buyers and sellers in a particular industry or industry segment and how those needs will be represented in the ANSI X12 standard. A document usually developed by an industry group, sometimes by a company, which describes the subset of a standard it intends to use to satisfy its EDI data requirements and the specific fields and codes that industry or company trading partners will use to pass information.

Hand-held (data entry) device: a portable piece of hardware on which data can be entered. It can be attached to other pieces of hardware such as a computer.

Handshaking: the required sequence of signals required to initiate and complete communication between system devices.

Hardware: the electric, electronic, and processing equipment used for processing data; any piece of automatic data-processing equipment, such as computer communication device, printer, or other computer peripheral.

Homogeneous data requirements: the uniformity of data elements required to transact business across company, industry, and country boundaries.

ICOPS: Intercompany Office Products Standard.

Imaging: placement of an optical representation of an object in an electronic data stream.

Incognito: with the real identity concealed.

Indirect benefits: benefits indirectly attributable to the implementation of EDI, usually requiring EDI plus system enhancements and procedural changes to be realized. Indirect benefits are typically larger than direct benefits but happen in the longer term and require a larger investment.

Industry-specific (implementation) guidelines: a document that describes the data needs of buyers and sellers in a particular industry or industry segment and how those needs will be represented in the ANSI X12 standard.

Industry-specific standard: an EDI standard that contains transaction sets and data fields that specifically satisfy the business needs of a particular industry group. Examples are UCS for the grocery industry, EAGLE for the hardlines industry, and ORDERNET/NWDA for the pharmaceutical industry.

Inelastic demand: demand that changes in relatively small volume in response to an increase or decrease in price.

Information float: the time period during which information is traveling from one place to another or is otherwise awaiting its next procedure.

Information-flow diagram: a visual representation of the flow of data throughout one or several functional business areas. Usually shows the information used during manual tasks, decision-making routines, and entry into a computer system.

In-network translation: the conversion of free-form data to an EDI standard or from one EDI standard to another by an EDI VAN.

Integrated applications: a series of computer programs designed such that the output of one can be automatically used in the next with no modifications or human intervention.

Interactive system: a computer program or programs with which a person can interact by inputting information via a computer terminal and receiving immediate visual feedback from the program.

Interchange: all data transmitted from one party to another in a single transmission. In an EDI standard, an interchange begins with an interchange header segment and ends with an interchange trailer segment.

Interchange envelope: the EDI standard entity that contains all the data transmitted from one EDI sender to one EDI receiver in a single transmission.

Interchange header (segment): an EDI standard segment which marks the beginning of an interchange. It contains the interchange identifier, sender and receiver IDs, and other data fields needed to interpret the incoming interchange.

Interchange level: refers to the receiving, sending, or processing of a full interchange.

Interchange trailer (segment): an EDI standard segment that marks the end of an interchange group. It contains the interchange identifier and the number of functional groups contained within the interchange.

Intercompany: pertaining to transactions crossing company boundaries.

Interconnect/interconnection: the transmission of EDI data from one EDI VAN to another via a communication link.

Internal edits: validation of data fields, either manually or as part of the logic of a computer program.

Internal file: a machine-readable computer file in which data are stored on a company site.

Internal format: see Application format.

Internal order: an order generated within a company and sent to its manufacturing or assembly location to fulfill an incoming purchase order or inventory requirements.

Internal staging order: an order to pick and accumulate or "stage" an order for shipment based on an incoming purchase order.

Internal systems: see Application Program.

Interpretation: refers to reading and understanding the meaning of EDI data that has been received from a trading partner.

Intracompany: pertaining to transactions within a company.

Inventory: a supply of goods kept on hand by a firm in order to meet needs either to fulfill customer orders, use in a manufacturing or assembling operation, or to respond to employee needs.

Inventory availability inquiry: an EDI transaction set requesting availability information on a particular product or products.

ISO: International Standards Organization.

IT: information technology—the department within a company that manages the computers and writes and maintains computer programs. Also called information services (IS) and management information services (MIS).

Invoice: a document prepared by the seller to apprise the buyer of the charge for product purchased and shipped by the seller.

Invoicing application: a computer program or group of programs that generates an invoice.

Item authorization transaction: a transaction set containing information on terms of sale and delivery schedule of product. Used in the direct store delivery segment of the grocery industry.

Item information transaction: an EDI standard transaction set that provides information regarding a product, for example, price, description, product code.

JEDI: Joint EDI Committee.

Just-in-time (JIT): a manufacturing philosophy that calls for receiving product at the manufacturing site just in time to be used in the next manufacturing process. Support of the just-in-time manufacturing philosophy requires very fast turnaround of orders and shipment of goods.

Just-in-time manufacturing schedules: a schedule of manufacturing tasks such that each task uses the output of the previous task in order to develop or assemble a complete product.

Keyed files: machine-readable computer files in which data are stored according to the value contained in one or more predefined key fields. This is done so data can be randomly accessed by searching for a particular value in a key field. For example, invoices may be stored by customer name, sales representative name, product purchased, and total dollar value.

Key entry: the entry of data into a computer by typing on the keyboard of a computer or computer terminal.

Key management: pertaining to the keys used to initiate authentication and encryption processes; the procurement and control of the keys required to perform these processes.

Labor-intensive: any process requiring a large proportion of human effort relative to capital investment.

Labor rates: level of wages.

LAN: local-area network—a connection of the computers of several business users within a department or office such that they may all access the same computer software or may have access to one another's computer files.

Levels of security: various degrees of access to system functions or data in order to limit ability to perform functions or to protect computer-stored data.

Line speed: see Bits per second.

Local area network: see LAN.

Lockbox: a post office box in which customer payments are deposited. The box is owned by a local bank who passes the payments to the company's main bank. A lockbox serves the purpose of reducing float for the receiver company by placing the receiving location closer to the customer.

Lockbox transaction: an EDI standard transaction set generated by a financial institution to describe payments received into a lockbox and lockbox balances.

Log on: a procedure used to access a system, involving exchange of information between user and system to identify the user to the system.

Log-on ID: the identification code used when logging on to a system, which permits the system to recognize the user. The log-on ID may be associated with a specific level of security and a predefined number of allowable tasks.

Looping: the repetition of a group of segments in an EDI standard transaction set. For example, all the segments that comprise a purchase order line item can loop (be repeated) up to 10,000 times.

Lot: a unit in the apparel industry made up of a predefined number of items in a predefined mix of color and size within style.

MAC: macroauthentication code—a character string computed as a result of the authentication of an EDI data stream. The MAC is appended to the original EDI data stream and transmitted along with the data. After performing the identical authentication computations at the receiving location, the receiver verifies that it has computed the identical MAC. If so, the receiver knows that the data are identical to those transmitted by the sender.

Machine-readable: in a format that is readily understandable by a computer program. Machine-readable files may be of fixed length, where each data field has a predefined length, or may be of variable length, where each data field has a minimum and maximum allowable length.

Mailbox: a logical partition of disk data storage in which all data sent to a particular recipient are stored until retrieved by that recipient. Mailboxes are provided as a service by EDI VANs.

Mail float: the time period after which a check is written and during which it, is in transit in the mail, and is presented to but not yet cleared by a bank.

Mainframe: technically, the fundamental portion of a computer, i.e., the portion that contains the CPU and control elements; commonly refers to any large computer.

Mainframe computer: generally characterized as being faster, more expensive, and having a larger set of capabilities and instructions than midrange computers or microcomputers.

Mandatory data element/segment: a data element or segment in an EDI standard in which information must be contained.

Manufacturing requirements: those components required to manufacture or assemble a product.

Map: the logical association of one set of values with values in another set. In EDI, moving data element values in an EDI data stream from one format to another, either between internal file and EDI standard or between one EDI standard and another.

Market factors: those situations outside of a company that influence its decision to implement EDI. For example, pressure from a major customer or competitive pressure.

Mass merchandiser: a large, low-level retailer, carrying a large variety of merchandise and competing for consumers mainly on the basis of price.

Material release: a transaction set used by the automotive industry which acts as an ordering mechanism for product covered by an annually negotiated purchase order.

Material requirements planning (MRP): a production scheduling process, usually computer-based, which arranges for the placement of materials and components where and when they are needed.

Menu: in an interactive computer system, a screen that is preformatted to display all the system functions from which a particular user may choose.

Menu-driven: refers to computer programs whose functionality can be accessed by a user via on-line, preformatted menu screens.

Message: a transmitted series of words or symbols that are intended to relay information. The EDIFACT standard refers to its electronic documents as messages.

Microcomputer: a general term referring to a complete, small computing system consisting of hardware and software—a personal computer (PC).

Midrange computer: a midsized computer, usually characterized by higher performance and higher price than microcomputers, having instructions sets, and a proliferation of high-level languages, operating systems, and network methodologies.

Model stock: a predefined collection of product styles, colors, and sizes that a particular retailer wishes to carry in a retail outlet.

MODEM: modulator/demodulator—an interface device that enables a computer to communicate digital information over telephone lines.

MRP: see Material Requirements Planning.

Multimedia: using more than one form of technology, for example, the transmission of data from a sender in an EDI standard format, conversion of that format by an EDI VAN to a print format, and transmission to the trading partner recipient via fax.

Multimedia EDI: passing data electronically via any combination of batch, interactive, print, and electronic mail media.

NACHA: see National Automated Clearing House Association.

NAK: see ACK.

National Automated Clearing House Association: a network of financial institutions through which financial transactions are cleared, crediting the payee and debiting the payer financial institution.

NCPD: National Counter Prescription Drug Plan.

NEX/UCS: network exchange of UCS data—refers to the EDI standard transaction sets used by the direct store delivery segment of the grocery industry; specifically those transaction sets transmitted between the supplier and the main office of the retailer to set up the EDI relationship and to share information regarding the products that will be delivered by the supplier and their prices.

Node: in a network, a point having both input and output circuits.

Noncustomer: the partner of a trading pair who sends and receives data through an EDI VAN but is not a customer of that VAN.

NRMA: National Retail Merchants Association.

NWDA: National Wholesale Druggists Association.

NWDA/ORDERNET: see ORDERNET/NWDA.

ODETTE: an EDI standard used by the European automotive industry.

OEM: original equipment manufacturer.

On-line: condition in which a user, terminal, or other device is actively connected with the facilities of a communications network or computer. Pertaining to a computer application, requiring input by a human.

Open network: a service provided by an EDI VAN to route data received from a VAN customer via autodial facility to a VAN noncustomer trading partner and to accept data from a VAN noncustomer and route it to the mailbox of the VAN customer.

Operational level manager: a midlevel manager responsible for the implementation of programs and completion of tasks designed to carry out strategic directives of the company.

Operational window: a predefined number of hours per day and days per week in which normal business can be transacted. In EDI, those hours when a company's data center is open and available to receive EDI data transmissions.

Operations staff: a group of employees who work in the data center of a company.

Optional data element: a data element of an EDI standard segment in which data can be, but does not need to be, present.

Optional data segment: a segment in an EDI standard transaction set which can be, but does not need to be, used.

Order-entry application: a computer program or group of programs which accepts and validates information from an incoming order and usually passes correct orders to an order-processing application.

ORDERNET/NWDA: an industry-specific standard designed for and used by wholesalers and manufacturers in the pharmaceutical industry. It was developed by the National Wholesale Druggists Association in conjunction with Informatics General Corporation.

Order-processing application: a computer program or group of programs that accepts orders validated by an order-entry application and further processes them by doing such processes as evaluating inventory availability, generating an internal staging order, and developing an internal order to a manufacturing or assembly location.

Order status inquiry: an EDI standard transaction set requesting information on the status of an order.

Outdial: see Autodial.

Paper-based business document: a business document whose information is printed on a paper form. Examples are purchase order, invoice, bill of lading.

Par level: the amount of inventory of a specific product that an organization would like to have resident at a predefined location.

Payment cancellation request: an EDI standard transaction set sent by a payer to its financial institution to cancel a previously sent payment instruction.

Payment from receipt (PFR): a methodology whereby a customer organization pays its vendor based on how much product it has received without the receipt of an invoice. The evaluated receipts settlement (ERS) used in the automotive industry is an example of a PFR system.

Payment order/remittance advice transaction: payment remittance transaction—an EDI standard transaction set that is used to authorize transfer of funds through a financial institution and/or to report on a payment.

Payment status report: an EDI standard transaction set sent by a bank to its customer to report on account balances and/or financial transactions.

Payment terms: see Discount terms.

PFR: see Payment from receipt.

Physical inventory: the actual counting of product to produce the total inventory-on-hand figure.

Pick instructions: the instructions to a warehouse person that describe what product to pick and how much, in order to stage a shipment.

PIDX: Petroleum industry data exchange.

Piggyback bar-code label: a bar-code label that contains two parts, both with the identical bar-code symbology. Typically one part is retained on the product, while the other is transferred to another location when the product is used. In hospitals, one-half of the piggyback is retained on the product packaging and the second half is affixed to a document that is used to update the patient's bill and degrade inventory-on-hand figures.

Planning schedule: an EDI standard transaction set used primarily by the automotive industry to describe a customer's forecasted manufacturing requirements.

Point of sale (POS): a computer application that permits the seller to collect product identification information at the time of sale, often picking up price information by cross-referencing the product identifier to the price file. Often a point-of-sale system will accept bar-coded product identification.

Potential payoff: see EDI Potential.

Predefined fixed file: a machine-readable computer file whose data elements and records are each of a predefined fixed length. For an EDI sender, this file is generated by the application link and used as input to the EDI translator. For an EDI receiver, this file is generated by the EDI translator and used as input by the application link program.

Preformatted screen: a computer terminal screen that has been designed to offer information or to accept entry into predesignated spaces.

Preordering transaction: one of several EDI standard transaction sets that contain information regarding requests for information and corresponding responses transacted during the preordering phase, for example, request for quote and the responding quote.

Price bracket: range of prices.

Price sales catalog data: an EDI standard transaction set that contains information to update an electronic catalog.

Print format: an organization of data that makes it easily human readable.

Product activity data transaction: an EDI standard transaction set that contains information on the sale, transfer, or return of product.

Product code: a code that uniquely identifies a product.

Product information transaction: an EDI standard transaction set that contains information to update product code, description, and other parameters.

Production mode: in data processing, automatic job scheduling and execution; occurs after programs have been debugged during test mode.

Program: see Application program.

Programmer: one who designs, writes, tests, and debugs computer programs.

Proprietary company standard: an EDI standard developed and owned by a company to handle transactions between that company and its trading partners. May be developed by a major customer or supplier organization.

Proprietary inquiry system: a program or group of programs developed and owned by a supplier, which is offered to its customers to respond to inventory and order status inquiries. Usually, such a system is interactive.

Proprietary ordering system: a program or group of programs developed and owned by a supplier, which is offered to its customers to handle ordering. Usually, such a system is interactive.

Proprietary standard: an EDI standard designed for and owned by a specific company to handle the information needs of that company.

Proprietary system: a program or group of programs owned by a company that handles specified processing for that company.

Protocol: a set of conventions between communicating devices that validates the format and content of messages to be exchanged and controls the communication session. Simple protocols define only hardware configuration; more complex protocols define timings, data formats, error detection, and correction technique

Public warehouse: a warehouse that is privately owned and is used to store product of many manufacturers.

Purchase order: a transaction set generated by a customer and containing all the information required by a supplier to process an order of product or services.

Purchase order acknowledgment: an EDI standard transaction set generated in response to a purchase order. It reports on availability of product and delivery details.

Purchase order change: an EDI standard transaction set containing original purchase order information and changes to that original order.

Purchase order change acknowledgment: an EDI standard transaction set generated in response to a purchase order change. It reports on availability of product and delivery details.

Purchase-pay cycle: period of time starting with generation of a purchase order, continuing through receipt of ordered product, and ending with payment.

Purchasing transaction: one of many transactions containing information relating to the purchase of a product or service, for example, purchase order, purchase order change, purchase order acknowledgment.

Qualifier code: a data element in which a code is placed that assists the receiver to interpret the meaning of information in a second, generic data element. For example, a qualifier code precedes the generic data element "Name." When the code SH is used, the receiver interprets the value contained in the name field to be Ship-to Name. Valid codes and their meanings as X12 qualifier codes are contained in code lists in the ANSI X12 data element dictionary.

Quick response (QR): a combination of technologies and business procedures and practices that eliminates waste and inefficiencies throughout the supply chain in the retail industry and maximizes responsiveness of vendors to their retailer customers.

Quote: an EDI standard transaction set that responds to a request for quote transaction set. It contains information on price, availability, and delivery schedule of a product.

Randomly accessed file: a data file whose records can be read in any sequence based on specified values of key fields.

Receiving advice: an EDI standard transaction set that reports on quantity and condition of received product.

Reconciliation procedure: a group of processing steps that validate and attempt to match information from various sources in order to discover matches and mismatches between the various data files.

Reengineering: the changing of systems and business procedures in order to streamline the transaction of business.

Relational database: a file of information in normalized form, i.e., a flat file that has been inverted.

Remote job entry: executing a program or group of programs by entering a series of commands to a computer from a remote source.

Request for quote (RFQ): an EDI standard transaction set containing information needed by a vendor to perform a request for price and availability quote for a product or service. Sometimes, this transaction set requires supporting documentation in the form of graphic designs, blueprints, and other visual aids to complete the required information.

Retrieval: in EDI, when a VAN customer communicates with the VAN, identifies itself, and requests transmission of data that have been sent to it by its trading partners.

RONW: return on net worth (also called return on equity)—a statistic that allows the management of a firm to analyze its ability to realize an adequate return on capital invested. Defined as net profit after taxes divided by net worth.

Safety stock: amount of inventory required to fulfill expected customer orders from the time that the purchase order is sent to the supplier for replenishment of stock until product arrives.

Scanner: an electronic device which converts the optical information from a bar code into electrical signals.

Scanning: examining information for a specific purpose as for content or arrangement through the use of an optical device.

Scanning device/scanner: an optical scanning unit which can read documents encoded in a special bar code.

SCM: see Shipment Container Marking.

Screen: see CRT.

Segment: in an EDI standard, a unit of a transaction set made up of related data elements.

Segment delimiter character: a control character that marks the end of data contained in a variable-length segment.

Segment diagram: in an EDI standard, the diagrammatic representation of a segment showing all its data elements.

Segment identifier: in an EDI standard, the code that uniquely identifies a segment, the segment identifier is the first data element in the segment.

Segment terminator: in an EDI standard, the control character that marks the end of data in a variable length segment.

Sequential file: a data file organized in such a way that its information must be accessed in sequential order from first to last record or segment.

Shipment container marking (SCM): the bar-coded label attached to the outside of a shipment container in order to identify the container and to provide additional information to the receiver such as final destination. The bar code used is represented in 128 bar-code symbology.

Shipment information transaction: an EDI standard transaction set that contains information regarding a shipment, sent by a carrier in response to a shipment inquiry transaction.

Shipment inquiry transaction: an EDI standard transaction set sent by a shipper or consignee (receiver) of a shipment that contains a request for information regarding a shipment in process.

Shipment staging application: a program or group of programs designed to use an internal staging order from which to develop pick instructions to fill an order.

Shipping transaction: one of several EDI standard transaction sets containing information relating to shipment of product.

Single sourcing: purchasing the full requirements of any one product from a single supplier.

SITPRO: British Simplification of International Trade Processing Board.

Smart card: a floppy disk which is used to carry EDI transactions from a direct store delivery supplier to the retailer. The smart card contains a copy of the invoice for the delivered product. It is inserted in the computer at the retailer's back door. If the invoice is correct, it is uploaded to the retailer's computer. If it is incorrect, it is corrected, uploaded to the retailer's computer, and down-loaded to the smart card. In either case, when the delivery is completed, both vendor and retailer have an identical copy of the correct invoice.

SMC: Standard Maintenance Committee—the group that handles maintenance of the UCS standard.

Software: any computer program.

Stand-alone: a microcomputer that acts as a complete unit containing application software, but not attached to any other computers.

Standard: in the ANSI X12 EDI standard, a transaction set that has been approved by the membership of X12 and has passed the public review process.

Standard business transaction: a set of business procedures between trading partners which accomplishes a specific purpose, for example, relaying the intent to purchase goods. Many transactions are characterized by a paper document which communicates the data required to accomplish the transaction, in this case, the purchase order. Electronic data interchange standard transaction sets are designed to accomplish this same goal without the use of a paper document.

Standard data format: a machine-readable file structure, predefined and agreed to by trading partners, which can accomplish a standard business transaction.

Standard syntax rules: rules governing structure of an EDI standard file.

Statement: an EDI standard transaction set sent by a vendor to its customer that contains a summary of the invoices sent during a predefined period of time such as a month.

Status details reply transaction: an EDI standard transaction set that is sent by a carrier in response to a shipment inquiry transaction set.

Stock items: items having predefined manufacturing and assembly specifications that are stocked by the vendor and can usually be described completely by a product code.

Stock-keeping unit (SKU): a number generated by a customer organization to uniquely identify a product. The Universal Product Code (UPC), a universally (by vendor and customer) accepted number, may be used as an SKU number.

Store and forward: pertaining to transmission of data, messages are received at an intermediate routing point and recorded (stored). They are subsequently transmitted to a further routing point or to the ultimate recipient.

Stow plan transaction: an EDI standard transaction set that contains information regarding the location where a specific shipment is stored in an ocean carrier.

Strategic manager: a high-level manager responsible for the long-range, strategic planning for his or her company.

Subset of a standard: a select group of transaction sets, segments, data elements, and qualifier codes of an EDI standard, such as ANSI X12. Some industry groups have defined the subset of the ANSI X12 standard that they will use.

Supplier base: vendor companies from which an organization purchases all its production and nonproduction products and services.

Supplier delinquencies: related to late shipments from a supplier.

Syntax rules of the EDI standard: rules governing structure and data usage of an EDI standard file.

TDCC (EDIA): Transportation Data Coordinating Committee (Electronic Data Interchange Association)—a company which has developed EDI standards for the transportation industry, the grocery industry, and the warehousing industry, and which has acted as the coordinating body of EDI standards for EDX, CIDX, and other groups.

Telecommunications: the transmission or reception of signals, writing, sounds, or intelligence of any nature by wire, radio, light beam, or any other electromagnetic means.

Telecommunication port: entry channel through with data are communicated. A port is part of a central computer system.

TERMS: a microbased software application developed by First National Bank of Chicago which allows the user to input pertinent parameters from which it computes a range of discount terms giving benefits to both partners of a trading partner pair.

Third party (service provider): EDI VAN—a company which acts as a communications intermediary between EDI trading partners, providing communications services such as line speed conversion, and protocol matching as well as electronic mailbox, in-network translation, carbon copy service, and other EDI services.

Trade organization: an association that represents the interests of buyer and/or seller companies of an industry or industry segment.

Trading partner: any company with whom an organization does business, for example, customer, supplier, carrier, bank.

Transaction level acknowledgment: a functional acknowledgment generated by the receiver of EDI data, which reports on receipt of a functional group and each transaction set in the group. A transaction level acknowledgment can report rejection of a single transaction set within a group.

Transaction (set): a collection of EDI standard segments that contain the information required by a receiver to perform a standard business transaction. In an EDI standard, a transaction set is defined as having three sections, header, detail, and summary, and is comprised of a predefined group of segments in each section.

Transaction set detail area: segments within the detail section of an EDI standard transaction set that contain information relating to the line items within the transaction set.

Transaction set diagram: a diagrammatic representation of the segments in a transaction set in the order in which they may be used.

Transaction set header area: segments within the header section of an EDI standard transaction set; they contain information relating to the entire transaction set.

Transaction set header segment: an EDI standard segment which marks the beginning of a transaction set.

Transaction set level: referring to the sending, receiving, or processing of data for a full transaction set.

Transaction set summary area: segments within the summary section of an EDI standard transaction set; they contain information relating to transaction set totals.

Transaction set trailer segment: a segment that marks the end of a transaction set.

Translation software: a program or group of programs that maps data between an application data format or a predefined fixed-field format and an EDI standard format.

Translator: see Translation software.

Transmission: an electronic transfer of a signal, message, or other form of intelligence from one location to another.

Transmission acknowledgment (TA): an ANSI X12 transaction set generated by a receiver of EDI data prior to validation by the translator and transmitted to the sender. It reports on receipt of a full transmission. This transaction set contains no report on validity or completeness of data.

Trigger: *noun*, any event that serves as a stimulus and initiates or precipitates a reaction or series of reactions; *verb*, to initiate or precipitate a reaction or series of reactions.

UCC: Uniform Code Council—an organization formed to administer the UPC symbol and other retail bar code and EDI standards. They are responsible for the assignment of UPC vendor numbers.

UCS: Uniform Communication Standard—an industry-specific EDI standard designed for and used by buyers and sellers in the grocery industry.

UCS/DSD task force: group of corporate members who participate on the direct store delivery task force, developing UCS transactions for this segment of the grocery industry.

UN/ECE: United Nations/Economic Commission of Europe—the organization which was given administrative control and responsibility for the EDIFACT standard.

UNIX: a multiprogramming operating system developed by Bell Laboratories that features sophisticated software and test developing utilities.

UNSM: United Nations standard message.

UOM: Unit of measure.

UPC: Universal Product Code—a code that uniquely identifies a product. It has three primary components. The first component is a six-digit code generated by the UCC that uniquely identifies the manufacturer of the product. The second is a six-digit code assigned by the manufacturer to uniquely identify each of its products. The third is a two-digit check number.

UPC-A: the name of the UPC number used to identify product in the retail and apparel industry. See UPC.

UPC catalog: a file containing UPCs and their associated parameter values.

Up-load: to transfer data from a front- or back-end processor to a larger computer.

Usage designator: associated with each data element within a segment, its usage characteristic, either mandatory, optional, or conditional. When conditional, the conditional rule is explained in a note.

Usage rules: see Guidelines for usage.

US Customs manifest: an EDI standard transaction set that contains information for the customs department regarding a shipment that is being imported.

User/business user: a member of the staff in a functional business area whose responsibility is to perform the tasks required to successfully fulfill the objectives of the business area.

User-friendly: easily understandable by a non–computer-oriented person.

User group: an organization made up of users of a specific system or product, it gives the users the opportunity to share knowledge they have gained and to identify additional system needs.

Value-added bank (VAB): a bank that acts as a network and communications intermediary.

VAN: value-added network—a company that acts as apipe for the transmission of data and provides communications services such as line speed conversion and protocol matching. See EDI VAN for additional features of a VAN.

Variable-length field: a data element whose length is determined by the amount of data contained in the field. Typically a variable-length field is defined as having a minimum and maximum length. For example, the name field may have from 1 to 35 characters in it. If we put the name Smith in it, we need only send the five meaningful characters in the name. All public EDI standards have variable-length fields.

Variable-length file: a file containing segments whose data elements are defined as variable-length fields. Such a file requires a data element delimiter character to mark the end of each data element and a segment terminator character to mark the end of each segment.

VAT: value-added tax.

Vendor-managed inventory (VMI): an arrangement whereby a vendor handles the replenishment of stock for a retailer. Usually done as part of a quick response system and is accompanied by a model stock plan previously agreed to by the vendor and retailer.

Vendor management: the evaluation of vendor performance based on preselected parameters such as quality of product and vendor performance.

Vendor performance: statistics for a specific vendor on actual delivery dates as compared to requested and promised dates and on product quality.

VICS: Voluntary Interindustry Communication Standard—a committee active in EDI on behalf of the apparel sector of the retail industry.

WAN: wide-area network—a connection of the computers of several business users within an organization such that they may all access the same computer software or may have access to one another's computer files.

WINS: Warehouse Information Network Standard.

WSA: Wholesale Stationers Association.

X12: see ANSI ASC X12.

X.25: communication protocol used in the X.400 environment.

X.400: CCITT standard by which electronic messaging is delivered. An X.400 header precedes the electronic message or messages being transmitted which allows the sender to specify information relating to the transmission and to specify delivery and notice requests.

X.435: a standard that further enhances the X.400 standard to make it deal more effectively with EDI transmission requirements.

X.500: addressing directory containing the names and characteristics of electronic messaging receivers. X.500 is used to facilitate the delivery of messages using X.400 and X.435 standards.

128 bar-code symbology: the particular bar-code symbology used when marking the outside of a shipping container with a bar code. Also called the UCC/EAN-128.

2780/3780 protocol: communication protocols that use or emulate IBM's binary synchronous protocol. 3780 can operate in an unattended mode.

Index

About the Author

Phyllis K. Sokol is currently Director of Educational Services at Sterling Software's Network Services Division in Columbus, Ohio. Prior to joining Sterling, she was president of EDI Consulting Services, a company she founded in 1984. As a consultant, she was retained by Fortune 500 companies during all phases of their EDI planning and implementation, where she specialized in working with functional business people to evaluate opportunities in EDI and establish an integrated EDI/business environmnent. Ms. Sokol has worked as an independent consultant and served on the EDI standards committee (ANSI) X12 since 1983. A popular speaker who has presented either workshops or keynote addresses at virtually all EDI-related conferences, she pioneered EDI for functional managers with the publication of the previous edition of this book, *EDI: The Competitive Edge*.